AN EVALUATION OF THE U.S. NAVY'S EXTREMELY LOW FREQUENCY COMMUNICATIONS SYSTEM ECOLOGICAL MONITORING PROGRAM

Committee to Evaluate the U.S. Navy's
Extremely Low Frequency Communications System
Ecological Monitoring Program

Board on Environmental Studies and Toxicology

Commission on Life Sciences

National Research Council

NATIONAL ACADEMY PRESS
WASHINGTON, D.C., 1997

NATIONAL ACADEMY PRESS 2101 Constitution Avenue, NW Washington, DC 20418

NOTICE: The project that is the subject of this report was approved by the Governing Board of the National Research Council, whose members are drawn from the councils of the National Academy of Sciences, the National Academy of Engineering, and the Institute of Medicine. The members of the committee responsible for the report were chosen for their special competences and with regard for appropriate balance.

This report has been reviewed by a group other than the authors according to procedures approved by a Report Review Committee consisting of members of the National Academy of Sciences, the National Academy of Engineering, and the Institute of Medicine.

The project was supported by contract DAMD 17-89-C-9086 between the National Academy of Sciences and the U.S. Department of Defense. Any opinions, findings, conclusions, or recommendations expressed in this publication are those of the author(s) and do not necessarily reflect the view of the organizations or agencies that provided support for this project.

Library of Congress Catalog Card Number 96-70733
International Standard Book Number 0-309-05590-3

Additional copies of this report are available from:

National Academy Press
2101 Constitution Ave., NW
Box 285
Washington, DC 20055
800-624-6242 or
202-334-3313 (in the Washington Metropolitan Area)
http://www.nap.edu

Copyright 1997 by the National Academy of Sciences. All rights reserved.

Printed in the United States of America.

COMMITTEE TO EVALUATE THE U.S. NAVY'S EXTREMELY LOW FREQUENCY COMMUNICATIONS SYSTEM ECOLOGICAL MONITORING PROGRAM

DUNCAN T. PATTEN *(Chair)*, Arizona State University, Tempe, Arizona
OM P. GANDHI, University of Utah, Salt Lake City, Utah
THOMAS GETTY, Michigan State Universy, Hickory Corners, Michigan
WILLIAM E. GORDON, Rice University, Houston, Texas
J. WOODLAND HASTINGS, Harvard University, Cambridge, Massachusetts
PETER KAREIVA, University of Washington, Seattle, Washington
JAMES C. LIN, University of Illinois, Chicago, Illinois
ROBERT G. OLSEN, Washington State University, Pullman, Washington
JOHN PASTOR, University of Minnesota, Duluth, Minnesota
BEVERLY J. RATHCKE, University of Michigan, Ann Arbor, Michigan
ANTONIO SASTRE, Midwest Research Institute, Kansas City, Missouri
LAWRENCE A. SHEPP, Rutgers University, New Brunswick, New Jersey and AT&T Bell Laboratories, Murray Hill, New Jersey

Staff

RAYMOND A. WASSEL, Project Director and Program Director for Environmental Sciences and Engineering
BONNIE A. SCARBOROUGH, Research Assistant
RUTH P. DANOFF, Senior Project Assistant
NORMAN GROSSBLATT, Editor
KATHRINE IVERSON, Information Specialist

Sponsor: U.S. Department of Defense

BOARD ON ENVIRONMENTAL STUDIES AND TOXICOLOGY

PAUL G. RISSER *(Chair)*, Oregon State University, Corvallis, Oregon
MAY R. BERENBAUM, University of Illinois, Urbana, Illinois
EULA BINGHAM, University of Cincinnati, Cincinnati, Ohio
PAUL BUSCH, Malcolm Pirnie, Inc., White Plains, New York
EDWIN H. CLARK II, Clean Sites, Inc., Alexandria, Virginia
ELLIS COWLING, North Carolina State University, Raleigh, North Carolina
GEORGE P. DASTON, The Procter & Gamble Co., Cincinnati, Ohio
PETER L. DEFUR, Virginia Commonwealth University, Richmond, Virginia
DAVID L. EATON, University of Washington, Seattle, Washington
DIANA FRECKMAN, Colorado State University, Ft. Collins, Colorado
ROBERT A. FROSCH, Harvard University, Cambridge, Massachusetts
DANIEL KREWSKI, Health & Welfare Canada, Ottawa, Ontario
RAYMOND C. LOEHR, The University of Texas, Austin, Texas
WARREN MUIR, Hampshire Research Institute, Alexandria, Virginia
GORDON ORIANS, University of Washington, Seattle, Washington
GEOFFREY PLACE, Hilton Head, South Carolina
BURTON H. SINGER, Princeton University, Princeton, New Jersey
MARGARET STRAND, Bayh, Connaughton and Malone, Washington, D.C.
BAILUS WALKER, JR., Howard University, Washington, D.C.
GERALD N. WOGAN, Massachusetts Institute of Technology, Cambridge, Massachusetts
TERRY F. YOSIE, Harrison/Ruder Finn Co., Washington, D.C.

Senior Staff Officers

JAMES J. REISA, Director
DAVID J. POLICANSKY, Associate Director and Program Director for Applied Ecology
KULBIR S. BAKSHI, Program Director for the Committee on Toxicology
CAROL A. MACZKA, Program Director for Toxicology and Risk Assessment
LEE R. PAULSON, Program Director for Information Systems and Statistics
RAYMOND A. WASSEL, Program Director for Environmental Sciences and Engineering

COMMISSION ON LIFE SCIENCES

THOMAS D. POLLARD *(Chair)*, The Salk Institute, La Jolla, California
FREDERICK R. ANDERSON, Cadwalader, Wickersham & Taft, Washington, D.C.
JOHN C. BAILAR III, University of Chicago, Chicago, Illinois
PAUL BERG, Stanford University School of Medicine, Stanford, California
JOHN E. BURRIS, Marine Biological Laboratory, Woods Hole, Massachusetts
SHARON L. DUNWOODY, University of Wisconsin, Madison, Wisconsin
URSULA W. GOODENOUGH, Washington University, St. Louis, Missouri
HENRY W. HEIKKINEN, University of Northern Colorado, Greeley, Colorado
HANS J. KENDE, Michigan State University, East Lansing, Michigan
SUSAN E. LEEMAN, Boston University School of Medicine, Boston, Massachusetts
THOMAS E. LOVEJOY, Smithsonian Institution, Washington, D.C.
DONALD R. MATTISON, University of Pittsburgh, Pittsburgh, Pennsylvania
JOSEPH E. MURRAY, Wellesley Hills, Massachusetts
EDWARD E. PENHOET, Chiron Corporation, Emeryville, California
EMIL A. PFITZER, Research Institute for Fragrance Materials, Hackensack, New Jersey
MALCOLM C. PIKE, University of Southern California, Los Angeles, California
HENRY C. PITOT III, University of Wisconsin, Madison, Wisconsin
JONATHAN M. SAMET, The Johns Hopkins University, Baltimore, Maryland
CHARLES F. STEVENS, The Salk Institute, La Jolla, California
JOHN L. VANDEBERG, Southwest Foundation for Biomedical Research, San Antonio, Texas

PAUL GILMAN, Executive Director

OTHER RECENT REPORTS OF THE BOARD ON ENVIRONMENTAL STUDIES AND TOXICOLOGY

Carcinogens and Anticarcinogens in the Human Diet: A Comparison of Naturally Occurring and Synthetic Substances (1996)
Upstream: Salmon and Society in the Pacific Northwest (1996)
Science and the Endangered Species Act (1995)
Wetlands: Characteristics and Boundaries (1995)
Biologic Markers (Urinary Toxicology (1995), Immunotoxicology (1992), Environmental Neurotoxicology (1992), Pulmonary Toxicology (1989), Reproductive Toxicology (1989))
Review of EPA's Environmental Monitoring and Assessment Program (three reports, 1994-1995)
Science and Judgment in Risk Assessment (1994)
Ranking Hazardous Waste Sites for Remedial Action (1994)
Pesticides in the Diets of Infants and Children (1993)
Issues in Risk Assessment (1993)
Setting Priorities for Land Conservation (1993)
Protecting Visibility in National Parks and Wilderness Areas (1993)
Dolphins and the Tuna Industry (1992)
Hazardous Materials on the Public Lands (1992)
Science and the National Parks (1992)
Animals as Sentinels of Environmental Health Hazards (1991)
Assessment of the U.S. Outer Continental Shelf Environmental Studies Program, Volumes I-IV (1991-1993)
Human Exposure Assessment for Airborne Pollutants (1991)
Monitoring Human Tissues for Toxic Substances (1991)
Rethinking the Ozone Problem in Urban and Regional Air Pollution (1991)
Decline of the Sea Turtles (1990)
Tracking Toxic Substances at Industrial Facilities (1990)

Copies of these reports may be ordered from
the National Academy Press
(800) 624-6242
(202) 334-3313

The National Academy of Sciences is a private, nonprofit, self-perpetuating society of distinguished scholars engaged in scientific and engineering research, dedicated to the furtherance of science and technology and to their use for the general welfare. Upon the authority of the charter granted to it by the Congress in 1863, the Academy has a mandate that requires it to advise the federal government on scientific and technical matters. Dr. Bruce Alberts is president of the National Academy of Sciences.

The National Academy of Engineering was established in 1964, under the charter of the National Academy of Sciences, as a parallel organization of outstanding engineers. It is autonomous in its administration and in the selection of its members, sharing with the National Academy of Sciences the responsibility for advising the federal government. The National Academy of Engineering also sponsors engineering programs aimed at meeting national needs, encourages education and research, and recognizes the superior achievements of engineers. Dr. William A. Wulf is president of the National Academy of Engineering.

The Institute of Medicine was established in 1970 by the National Academy of Sciences to secure the services of eminent members of appropriate professions in the examination of policy matters pertaining to the health of the public. The Institute acts under the responsibility given to the National Academy of Sciences by its congressional charter to be an adviser to the federal government and, upon its own initiative, to identify issues of medical care, research, and education. Dr. Kenneth I. Shine is president of the Institute of Medicine.

The National Research Council was organized by the National Academy of Sciences in 1916 to associate the broad community of science and technology with the Academy's purposes of furthering knowledge and advising the federal government. Functioning in accordance with general policies determined by the Academy, the Council has become the principal operating agency of both the National Academy of Sciences and the National Academy of Engineering in providing services to the government, the public, and the scientific and engineering communities. The Council is administered jointly by both Academies and the Institute of Medicine. Dr. Bruce Alberts and Dr. William A. Wulf are chairman and vice chairman, respectively, of the National Research Council.

Preface

THE U.S. NAVY has an extremely-low-frequency (ELF) communications system that transmits from Wisconsin and upper Michigan using electric and magnetic fields (EMFs) to communicate with submarines anywhere in the world. In response to recommendations in a 1977 National Research Council report, the Navy conducted a multiyear program to monitor possible effects on plants and animals in the vicinity of the transmitting antennas. Possible effects on humans were not addressed in the program.

The Navy arranged for the Illinois Institute of Technology Research Institute (IITRI) to manage the ecological monitoring program and provide engineering support, such as performing ELF-EMF measurements. The program included 11 monitoring studies, which were undertaken by subcontracting researchers from Michigan State University, Michigan Technological University, the University of Minnesota-Duluth, the University of Wisconsin-Milwaukee, and the University of Wisconsin-Parkside. Each research team prepared a final report on the results and conclusions of its multiyear effort.

In 1995, at the Navy's request, the National Research Council convened the Committee to Evaluate the U.S. Navy's Extremely Low Frequency Communications System Ecological Monitoring Program to review independently the results of the Navy's multiyear program. The committee was not charged or constituted to address the broader topic of biologic effects of EMFs. Thus, it did not separately investigate mechanisms by which EMFs might affect biologic systems. Rather, the committee placed its emphasis on reviewing and evaluating the study designs, analyses of data, and interpretations of results in the Navy's program. Possible human effects of EMF exposure have been addressed in another National Research Council report, *Possible Health Effects of Exposure to Residential Electric and Magnetic Fields* (NRC 1997).

As part of its analysis, the committee reviewed documents from the ecological monitoring program, including the requests for proposals, the proposals themselves, annual and final reports, reviewer comments, and engineering reports. The committee also received information in response to questions that it asked of the Navy, IITRI, and subcontracting researchers. Documents from the Ecological Monitoring Program are available to the public from the National Technical Information Service, Springfield, Va.

The committee sought additional levant information from grass-roots organizations active in protesting the use of the ELF system and from various individuals with scientific expertise on the effects of electric and magnetic radiation on plants and animals. No data relevant to ecological effects were obtained through such requests.

At its first meeting in July 1995, the committee heard presentations by Dennis Murphy, Communications Systems Project Office, Space and Naval Warfare Systems Command, U.S. Navy; Anthony Valentino, Vice President, IITRI; John Zapotosky, Program Manager, IITRI; and Abdul El-Shaarawi, statistical consultant to IITRI. Information was also provided by Robert Yacovissi, Non Ionizing Radiation Health Branch, Bureau of Medicine and Surgery, U.S. Navy; Willie Jones, Space and Naval Warfare Systems Command, U.S. Navy; Bonnie Bonner, Space and Naval Warfare Systems Command, U.S. Navy; and Ralph Carlson, Director of Research, ELF Electromagnetic Compatibility Assurancy, IITRI.

The committee thanks all the persons mentioned above. In addition to his presentation, John Zapatosky provided helpful and timely information to the committee over the course of its deliberations. The committee also acknowledges the efforts of individual researchers in the monitoring program who generously responded to requests for information.

Although this report represents the work of the committee, it benefited from the excellent contributions of staff at the National Research Council: Ray Wassel, Bonnie Scarborough, Ruth Danoff, and Larry Toburen. Norman Grossblatt edited the report.

—Duncan T. Patten, Chair

Contents

	EXECUTIVE SUMMARY	1
1	INTRODUCTION	13

    The ELF Communications System, 13
    The Navy's Ecological Monitoring Program, 15
    The Committee's Charge and Approach, 17
    Specific Theories of Biologic Effects of EMF Exposure, 19
    Scope of the Report, 20

2 EMF MEASUREMENT, EXPOSURE CRITERIA, AND DOSIMETRY 21
    Characterization of Electric and Magnetic Fields, 22
    Exposure Criteria for Site Selection, 25
    Exposure Data Supplied to Researchers, 26
    Using Formulas for Predicting Electric and Magnetic Fields, 26
    Dosimetry, 28
    Differences in Effect Between Unmodulated 60-Hz and Modulated 76-Hz Signals, 30
    Conclusions Regarding EMF Measurements, 30

3 EVALUATION OF FINAL REPORTS OF INDIVIDUAL STUDIES 32
    Introduction, 32
    Wetlands, 33
    Slime Mold, 45

Wisconsin Birds and Michigan Birds, 53
Small Vertebrates, 61
Litter Decomposition and Microflora, 73
Upland Flora, 85
Aquatic Ecosystems, 92
Pollinating Insects, 96
Soil Arthropods and Earthworms, 104
Soil Amebas, 108

4 **COMMON ISSUES**   111
Use of Exposure Data by Ecological Monitoring Teams, 111
Study-Site Selection, 116
Pseudoreplication, 118
Species Selection, 120
Response-Variable Selection, 122
Statistical Power, 126
Response to Reviews and Critiques, 128
Appropriateness of Interpretation, 130
Different Methods for Similar Organisms, 131
Lack of Integration Among Studies and Synthesis of Information, 132
Data Archiving, 135

5 **OVERALL CONCLUSIONS AND RECOMMENDATIONS**   138
Ecological Effects, 139
IITRI's Engineering Support and Program Management, 142
Recommendation, 143
Suggested Next Steps, 144

References   148

Appendix A   153

Appendix B   157

AN EVALUATION OF THE U.S. NAVY'S EXTREMELY LOW FREQUENCY COMMUNICATIONS SYSTEM ECOLOGICAL MONITORING PROGRAM

Executive Summary

THE U.S. NAVY'S EXTREMELY-LOW-FREQUENCY (ELF) communications system consists of a transmitting facility in northern Wisconsin, near Clam Lake, and another facility in Michigan's Upper Peninsula, near Republic. In 1982, the Navy established an ecological monitoring program to determine whether electric and magnetic fields (EMFs) from the ELF communications system affected plant and animal populations or otherwise caused ecological changes in the areas surrounding the transmitting facilities.

The ecological monitoring program comprised 11 studies of wetlands, slime mold, birds, small vertebrates, litter decomposition and microflora, upland flora, aquatic ecosystems, pollinating insects, soil arthropods and earthworms, and soil amebas. The Illinois Institute of Technology Research Institute (IITRI) provided management and engineering support for the program. Ecological studies were carried out by academic researchers. The monitoring program was completed in 1995.

Each study investigated possible effects of EMFs by taking biologic measurements at a location (referred to as a treatment site) near one of the ELF transmitting facilities and at a corresponding control site where EMF strength was no more than one-tenth of the EMF strength at the treatment site. Most studies also addressed differences in factors other than EMFs from the ELF antenna, such as soil moisture. Measurements taken over time from pairs of treatment and control sites were analyzed statistically to test for significant differences between the sites.

To summarize the 11 final reports of biologic research: IITRI reported that although some of the program's researchers believe that a few biologic changes might have occurred, all stated that there were no consistent, unequivocal effects of ELF antenna operation on any of the variables they monitored. All concluded that the implications of their results do not indicate adverse ecological effects of significance due to the ELF facilities.

At the request of the Navy, the National Research Council formed the Committee to Evaluate the U.S. Navy's Extremely Low Frequency Communications System Ecological Monitoring Program in 1995 to evaluate independently the program's objectives and design, data-collection methods, data analysis, and interpretations. The studies included in the ecological monitoring program were evaluated with standard criteria for all branches of scientific endeavor, including appropriateness and coherence of the hypotheses being tested, adequacy of experimental design to test these hypotheses, methods for data collection and analysis, and the soundness of conclusions drawn from investigators' observations.

Throughout its review, the committee was aware of controversies surrounding scientific theories regarding the manner in which biologic systems might be affected by EMF exposure. However, it was beyond the committee's mandate to address more general questions of the plausibility of different theories of biologic effects or the plausibility of there being such effects at all. The Navy's studies were evaluated according to the criteria mentioned above—not for their ability to confirm or disprove specific theories of biologic effects. Also, the committee was not asked to determine whether EMFs in general (that is, from all sources) are a matter of concern. Rather, it reviewed the ecological monitoring program's assessment of the possible effects of the operation of the ELF communication system.

COMMON ISSUES ARISING FROM EVALUATION OF INDIVIDUAL STUDIES

The committee's evaluation of the 11 ecological studies in the Navy's ELF ecological monitoring program revealed several issues that were common to many or all of the studies. Those issues are summarized below.

USE OF EXPOSURE DATA BY ECOLOGICAL MONITORING TEAMS

Measurements of electric and magnetic fields in the vicinity of the study sites were made by IITRI personnel at regular intervals, and they provided this information to researchers. IITRI also provided operating records of the transmitting facilities. Such records were needed because the facilities were not operated continuously, even during years when they were fully operational, so study sites were not consistently exposed to ELF EMFs. At some times, different portions of the antennas were turned on and off several times each day, with varying modulations, frequencies, current intensities, and phase angles. In a number of studies, especially those with short-term response variables, it was important to know whether the transmitter was on or off during critical exposure periods. However, there is little evidence that the research teams, except the upland flora and wetlands teams, considered or were aware of this factor in evaluating the results of their experiments.

STUDY-SITE SELECTION

Site selection was not easy. So many conditions had to be met that perfect matches between treatment sites and control sites were impossible. The program would have been improved by studying fewer variables at more and larger sites and by eliminating studies with poorly matched treatment and control sites. For example, in the soil arthropods and earthworms study, the dominant earthworm studied at the treatment site was not found at the control site. This experiment could never yield results that could clearly be attributed to ELF-EMF exposure, because there was no way of separating ELF-EMF effects from other factors without a control site at which the same earthworm was studied.

ADEQUACY OF SITE REPLICATION

In many of the studies, the main effect of interest, namely the effect due to the presence of ELF EMFs produced by the antennas, was pseudoreplicated (that is, not truly replicated) because there was only one study site for each level of exposure. The experimental data therefore only provide an estimate of the variance of responses studied within each site. They do not provide an estimate of the variance due to exposures across sites. The effects of the antennas on response variables are therefore confounded with the background effects of the different soils, climate, and other characteristics of each site.

A danger here is accepting the null hypothesis (that is, no effect of exposure to ELF EMFs) when it might be, in fact, false. In many of the studies, it is not possible to calculate the probability of such an error, because it depends on an independent estimate of differences between treatment and control, which requires replication of sites, not simply of plots within sites. The acceptance of all further conclusions must proceed with these caveats in mind, but the caveats are generally not stated clearly anywhere in the reports.

SPECIES SELECTION

The diversity of species studied was considerable. It included representatives of most major taxonomic groups and types of organisms that have been reported to show effects of ELF-EMF exposure in previous laboratory or field studies.

Aspects of concern include the lack of studies of amphibians and nonvascular plants. In the wetlands study, a moss population was found to increase significantly at the treatment site, but the response was not pursued, because the moss was not a target species; this is unfortunate because the finding might be an indicator that moss is especially sensitive to ELF-EMF effects.

No studies focused on rare species or potentially endangered species. These omissions are problematic. Some rare species, such as predators and keystone species, can exert major effects on communities. It is not known whether rare populations at the edge of their range are more or less sensitive to bioassays of additional stresses than abundant species.

RESPONSE VARIABLES

Many short- and long-term response variables were measured, including characteristics of bird populations, soil microbiology and ecology, plant ecology, insect populations and behavior, water quality, fish ecology, and reproduction. Such breadth is commendable for detecting potential effects at different ecological levels. However, the ability of the studies to detect possible ELF-EMF effects was generally weak. Antenna on-off activity was not related to measurements of short-term biologic responses. Possible small effects could have been difficult to detect, given the lack of statistical power of some studies and the occurrence of confounding variables. The term "small effects" is used in this report to refer to ecological effects whose magnitudes are not likely to exceed those expected from normal perturbations over the short term. A drought is one example of such normal perturbations. The few small possible effects that were found were often too readily dismissed in the study reports. The lack of knowledge of possible mechanisms in the scientific community and the lack of use of simulation models by the ELF researchers might have reduced the researchers' ability to decide which organisms and response variables are most likely to exhibit effects of the ELF antenna.

STATISTICAL POWER

The statistical power of an experimental design reflects the likelihood that an experiment will be able to detect the presence or absence of a phenomenon being studied. The committee noted lack of adequate statistical power as a problem that arose in more than one study. Studies with low statistical power would not be able to accept or reject the null hypothesis with sufficient confidence. For example, in several components of the small vertebrates studies, the statistical power was dropped from 90% to 70%, thereby making it much less likely that effects, if any were present, would be detected. Some design changes led to substudies with statistical power of 30%. Such experiments offer no bounds on uncertainty, and it can reasonably be questioned whether such experiments should have been performed at all.

RESPONSE TO REVIEW

There was a great deal of variation in how IITRI and the researchers responded to external reviews. In many cases, the investigators were very responsive to reviewers' comments and critiques; that contributed greatly to their research and to the results of the monitoring program. In some cases, reviewers' comments were addressed through explanations of why suggestions were not heeded. In a few cases, suggestions were not followed, and research design, analytic techniques, or interpretations that needed improvement or correction were left unattended. Some of the reviewers' comments were very critical regarding statistical power, data archiving, and unwarranted dismissal of possible effects. Such criticisms should have been a clarion call to program managers that some studies had substantial problems.

APPROPRIATENESS OF INTERPRETATION

In several studies, modest but significant differences were observed between data collected at treatment sites and data from control sites. Researchers conducting the studies concluded that five of these potential effects were due to factors other than the ELF antenna. Without attempting to judge whether any of those interpretations suggested a predisposition to a particular outcome, it is important to consider whether the conclusions were established with a credible scientific basis. In the course of this committee's review and discussion of the researchers' final reports, concerns arose about the scientific credibility of some of the conclusions.

Differences between treatment sites and control sites that were dismissed by researchers and by IITRI as not being clearly related to ELF exposure included the increase in bee overwintering mortality, the reduction in leaves per bee nest cell, accelerated litter decomposition, early eye-opening in mice, and depressed earthworm reproductive rates. The committee believes that some of those differences were dismissed too readily as alleged artifacts of environmental variations or experiment design.

DIFFERENT METHODS FOR SIMILAR ORGANISMS

The broad range of studies in the program often resulted in more than one research team's examining possible ELF-EMF effects on similar organisms or processes, but with somewhat different protocols. The use of different methods is sometimes advantageous or necessary and does not necessarily negate conclusions from any one study. However, the use of different methods prevented the researchers from making valid cross-site comparisons and thereby impeded the realization of the full potential of integration across sites and organisms. There does not appear to have been much discussion of coordinating methods before the experiments began.

LACK OF INTEGRATION AMONG STUDIES AND SYNTHESIS OF INFORMATION

Early research design should have been guided by recognition of interactions among ecosystem components and by encouragement of integration among studies with full development and application of appropriate statistical approaches. If there had been consistent integration and comparison of findings of the different studies, the overall research effort could have been improved with the same expenditure of resources. With better integration, there could have been more pursuit of promising results, the hallmark of good research. By failing to integrate the studies of different species and ecosystem processes, this large-scale effort largely surrendered the possibility of detecting small changes in interactions of components and gave up the major advantage of such large-scale research.

DATA ARCHIVING

There appears to have been no standard procedure for archiving data resulting from the ecological monitoring program. For some studies, original data were kept primarily in laboratory notebooks. As far as the present committee is aware, no standard protocols were established for formatting, documenting, or reporting the data. Most of the data were never transferred to the program manager, and those data are not readily available. As program manager, IITRI should have been responsible for data archiving and planning for long-term availability of the information.

IITRI'S ENGINEERING SUPPORT AND PROGRAM MANAGEMENT

IITRI did a good job on the engineering aspects of the ecological monitoring program in characterizing the spatial and temporal characteristics of the electric and magnetic fields. The instrumentation for ELF-EMF measurements appeared to be well designed, well calibrated, and properly used. IITRI provided ELF-EMF exposure information to the researchers for each study. In addition, IITRI was responsive to requests from researchers for additional engineering support.

However, in its review of individual study reports, as well as of the overall program of monitoring for possible effects of the ELF antennas, this committee discovered weaknesses in some aspects of IITRI management of the program. Three of these weaknesses appear to have been caused by lack of adequate oversight. First, IITRI should have detected problems with the use of exposure data through annual research reports. In response, it should have provided more guidance for use of the exposure information and required that an EMF-exposure expert work closely with each study until the study leader understood the types of data available, data variability, and the best methods for applying the data. And there should have been greater involvement of an expert in broad-based applied statistics at the earliest phases in the design of this program's studies. Second, responsibility for the lack of archiving and of planning for long-term availability of monitoring-program information appears to rest with IITRI's management of the program. Third, IITRI should have established a regular internal-review process to ensure that each study adequately addressed external criticism. Those problems appear to have originated in poor early planning by IITRI or inadequate followup by IITRI as problems arose during the duration of the program.

ECOLOGICAL EFFECTS

The Navy's ecological monitoring program reported no obvious adverse ecological effects, such as unusual changes in species populations or large-scale mortality of trees or other organisms, as a result of operation of the ELF communications system during the period of the monitoring program. The monitoring program also did not detect any small effects with well-defined consequences, such as decreases in reproductive fitness, that would be likely to result in major effects in the future.

The present committee agrees with the general findings of the Navy's ecological monitoring program, within the limitations described in this report, that the researchers' observations provide no evidence of statistically significant, widespread, adverse effects of EMFs associated with the ELF antennas on bird populations, leaf-litter decomposition processes, upland flora, the movements of dragonfly larvae, the colonization of leaf litter, the movement of fish, soil arthropod populations, and soil ameba populations. However, some of the studies, as discussed in this report, had deficiencies that diminished their capabilities to detect small effects.

The committee recognizes that small effects on populations, mediated through modest changes in response variables, might slowly compound and only later become apparent. Numerous flaws in the ecological studies—as designed, implemented, and interpreted—would have compromised detection of many possible small effects of the antenna operation. As discussed in Chapter 5, the individual ecological studies can be sorted into three categories: those which the committee judged to be acceptable with qualification, those which might be acceptable after more information is obtained or data are reanalyzed, and those which are unsalvageable because of serious flaws.

RECOMMENDATION

The complexity of assessing possible ecological impacts of ELF EMFs—especially given the diversity of the ecosystems and the variability of their locations and their distances from the antennas, (and therefore the variability of exposures)—made it extremely difficult to design and complete appropriate and comparable ecological monitoring studies. The following recommendation and suggested next steps reflect this committee's understanding of such difficulties but also indicate the committee's concern for bringing this ecological monitoring program to an appropriate and fruitful conclusion.

DO NOT REPEAT THE FIELD STUDIES

Dispite the weaknesses of the monitoring studies, the committee does not recommend that the field studies be repeated, because the extensive studies conducted to date have provided no evidence that exposure to ELF EMFs had obvious adverse ecological effects. Although caution must be used in drawing conclusions from results of most of the studies regarding possible small effects because of faulty study design or analysis, the committee considers it highly unlikely that repetition of the ecological monitoring studies undertaken in this program would produce any new findings about ecological effects of ELF EMFs.

SUGGESTED NEXT STEPS

REANALYZE THE EXPOSURE-ASSESSMENT DATA

The ELF ecological monitoring studies were supplied with ELF-EMF data based on measurements made by IITRI engineering teams. The timing and location of the measurements differed among studies. They included measurements made only once a year at some study plots (as in the wetlands study), at the location of each individual of the response species of interest in others (as in the upland-flora study), and as a spatial gradient of exposure levels (as in the bird-nestling study). In addition to those different forms of available ELF-EMF exposure data, the study teams apparently were made aware of the variability in times and outputs of antenna operations. In some studies, the analyses and interpretations of ELF-EMF effects appear to have made use of the exposure data. However, some studies apparently used the information inappropriately, and others might not have fully recognized the importance of the vagaries of antenna operations and output.

The committee suggests that the investigators from each ecological monitoring study reassess their use of data on antenna operation and ELF-EMF data and, if they were used inappropriately, reanalyze the responses of selected ecological variables. In addition, the results of several studies should be reanalyzed so that firmer conclusions can be drawn. These are the studies labeled in Table 5-1 as "Might Be Acceptable With More Information or Analysis." The committee suggests that an organization that is independent of the Department of Defense and IITRI direct the reanalysis. The reassessment and reanalysis should be performed in close collaboration with biostatisticians familiar with this type of EMF-exposure assessment and engineers knowledgeable about field ELF-EMF exposure measurements. If reanalysis reveals statistically significant or suggestive responses of ecological variables to ELF EMFs, these responses could be considered for further controlled study (as discussed below). The committee suggests reanalysis and possible controlled studies so that an opportunity to improve the understanding of ELF-EMF exposure and possible ecological responses is not lost.

VARIABLES THAT TENDED TO SHOW MEASURABLE EFFECTS SHOULD BE SUBJECTED TO CONTROLLED LABORATORY STUDY

The ELF ecological monitoring studies produced few results that tended to show effects of ELF-EMF exposure on selected ecological variables. One effect, the growth response of upland trees, might be an artifact of selective use of exposure data. In other cases, an effect might be a true measurable response, but the experimental design or the complexity of the surrounding ecosystem might have created an environment that made the findings sufficiently questionable to warrant further, more-controlled studies.

Responses that perhaps could be tested under controlled laboratory conditions are the apparent increase in chlorophyll-a in the aquatic-ecosystem study, the behavioral responses of bees and their overwintering mortality in the pollinating-insects study, and increased moss growth in the litter-decomposition study. The chlorophyll-a increase appeared to be an increase in cell density rather than in chlorophyll per cell; this possibility could be tested and the ecological implications analyzed. A similar study might help in understanding whether alterations in bee behavior and mortality are repeatable and can be shown to be caused by ELF-EMF exposure or are artifacts of the less-controlled, more-complex study sites. The wetland study unexpectedly discovered more moss cover on decomposition bags closer to the antenna treatment sites than in intermediate treatment or background control sites. The increased moss cover caused problems in interpreting data on decomposition, but the variability in growth of moss should be considered for controlled investigation. The reanalysis of exposure assessments (as discussed above) might uncover additional suggestions of small but measurable responses of ecological attributes to ELF-EMF exposure. If it does, these responses could also receive further study under more-controlled conditions; such studies could be designed also to help to elucidate the mechanisms of response if an EMF effect is observed. Such information might guide researchers in deciding which organisms and response variables are most likely to exhibit effects, if any, of the ELF antenna.

REANALYSIS OR LABORATORY STUDIES SHOULD BE REVIEWED INDEPENDENTLY

Reanalysis of exposure assessments might or might not identify some effects of ELF-EMF exposure on ecological variables not previously observed, and laboratory tests might or might not confirm them. Reanalysis might also strengthen the credibility of the findings of some studies. The committee suggests that if reanalyses or laboratory studies are performed, the Navy should arrange for an independent evaluation by a few individuals to assess all of the findings resulting from the reanalyses. The individuals should include biostatisticians familiar with ELF-EMF exposure assessment and biologic expertise to determine what the weight of evidence indicates and the biologic or ecological implications of any substantiated treatment effects. A broader integration of all studies should be pursued through the use of quantitative methods designed for such purposes. Integration of related effects, although not statistically significant, can point to areas where additional study might be warranted. The results of the independent evaluation should be made publicly available. Such an independent final review would serve the Navy and the public in producing more-credible and improved findings of the monitoring program.

1

Introduction

THE ELF COMMUNICATIONS SYSTEM

IN THE LATE 1950s, the U.S. Navy began to investigate the use of extremely-low-frequency (ELF) electric and magnetic fields (EMFs) to communicate with submerged submarines. In 1969, the Navy completed an experimental facility, in the Chequamegon National Forest near Clam Lake, Wisconsin. For more than a decade, this facility was operated intermittently and at less than full power to test the system and to make engineering evaluations. The Naval Radio Transmitting Facility (NRTF) in Wisconsin was upgraded and became fully functional in 1985. About 150 miles to the east—in the Copper County and Escanaba River state forests near Republic, Michigan—another facility was completed by the Navy in 1986. It was operated intermittently beginning in 1986 and became fully operational in 1989.

Today, the Navy's ELF communications system comprises the radio transmitting facility near Clam Lake, Wisconsin, and the one near Republic, Michigan (see Figure 1-1). Each facility consists of a transmitter connected to long overhead wires (antennas) with ground terminals buried at their ends. The antennas and terminals are in cleared rights of way in forests. The rights of way are 70 to 100 feet wide. The Wisconsin facility has two antennas, each 14 miles long, perpendicular to each other. The Michigan facility has three antennas: two are 14 miles long and parallel to each other, and the third is 28 miles long and perpendicular to the other two.

14

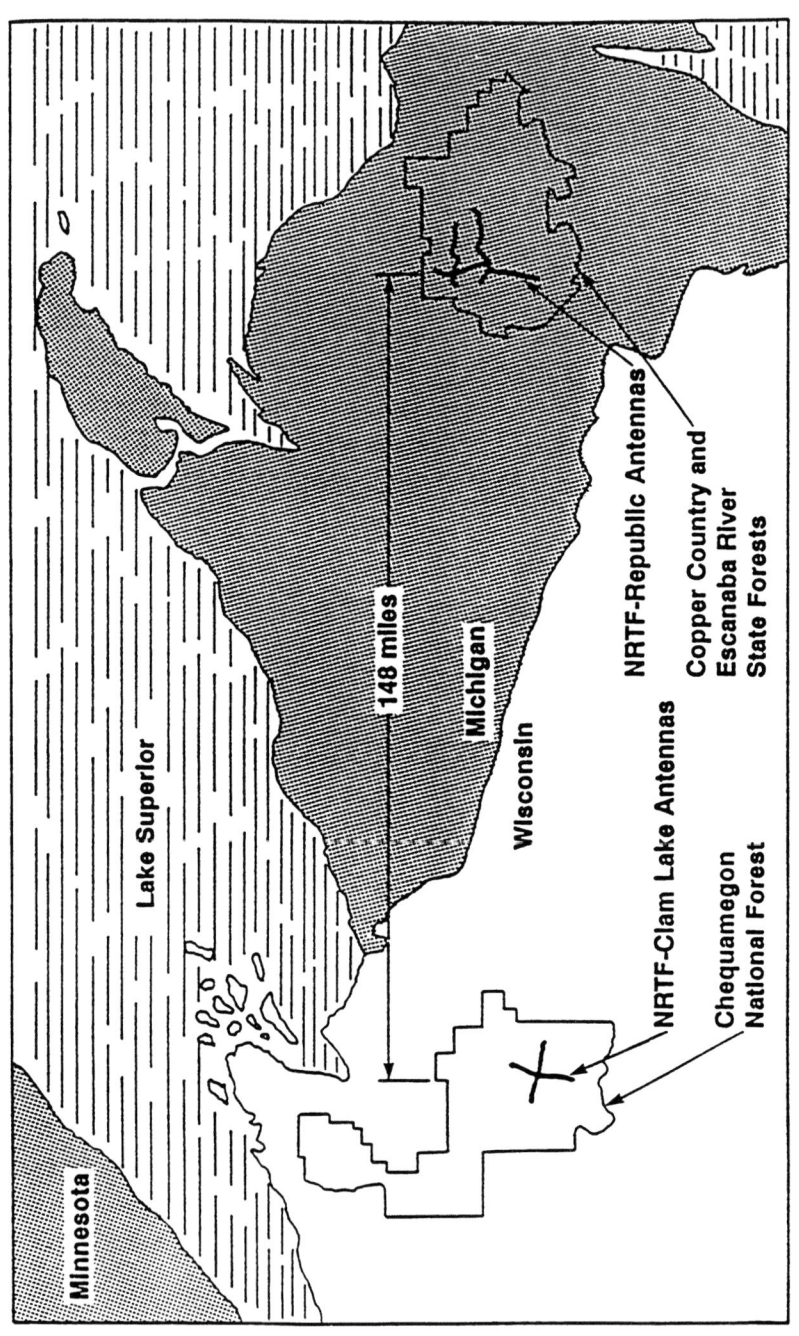

FIGURE 1-1 ELF communications facilities in Wisconsin and Michigan. Source: Haradem et al. 1993.

The frequency of the EMFs produced by the ELF communications system is modulated between 72 and 80 hertz (Hz) to produce a binary-coded signal (1 or 0) that is transmitted to submarines. The most prominent frequency is 76 Hz (Haradem et al. 1994). The transmitters use three ELF EMFs to broadcast messages to submarines:

- A magnetic field in the air and the earth that is generated by the electric current in the antennas and ground terminals.
- An electric field in the earth that results from the combination of fields induced by the magnetic field and the current flowing from the buried ground terminals.
- An electric field in the air that is produced as a result of the difference in electric potential between the antennas and the earth or as a byproduct of the electric field in the earth.

THE NAVY'S ECOLOGICAL MONITORING PROGRAM

At the time of the construction of the Wisconsin experimental transmitter, the Navy provided funds for the U.S. Forest Service and other investigators to determine whether changes in biologic factors could be observed in response to electric and magnetic fields produced by the existing Wisconsin experimental facility. In 1977, the National Research Council independently evaluated the studies at the Wisconsin facility and the results of other studies on effects of ELF EMFs (NRC 1977). The Navy undertook a similar exercise. Both the Navy and the Research Council concluded that substantial effects on organisms in the vicinity of the facility would be unlikely, but recommended establishment of a biologic monitoring program at the ELF communications facilities if they became operational. Years later, the American Institute of Biological Sciences (AIBS) reviewed bioelectromagnetic research and recommended continued ecological research because of the problems of extrapolating results of highly controlled laboratory studies to complex ecological systems (AIBS 1985).

The original plan for the ecological monitoring program was developed by the Navy in the late 1970s to satisfy such recommendations. This program included a set of general, statistical, and design requirements. It also required ambient environmental monitoring and the development of a preconstruction database. The ecological studies listed included soil microfauna, earthworms, small mammals, large mammals, crops, herbs, grasses, trees, periphytic algae, aquatic insects, fish, pollinating insects, nesting birds, migrating birds in flight, and human health effects. Data were to be interpreted by principal investigators, and the monitoring program was to be supervised by a review committee composed of Navy and state representatives.

The Illinois Institute of Technology Research Institute (IITRI) received the contract to manage the ecological monitoring program and provide overall program technical support. IITRI was in a unique position because of its long experience of technical support to the ELF project, including work to ensure that the ELF communications system did not interfere with local power distribution lines or telephone circuits. IITRI had had previous responsibilities for environmental design, protection planning, and engineering support for biologic studies. Under the ecological monitoring program, IITRI's management roles were program coordination, study evaluation and monitoring, subcontract administration, engineering support, and direction of a review of data interpretation and reporting.

IITRI designed a fully developed ecological monitoring program in 1982 based on the original outline from the Navy, input from several state and federal agencies, the 1977 National Research Council report, and the Navy's draft environmental-impact statement. The purpose of the ecological monitoring program was to determine whether EMFs produced by the ELF communications system affected plant or animal populations or otherwise caused important ecological community or ecosystem changes. The program was included in a 1982 request for proposals (RFP) to perform monitoring studies.

The RFP stipulated that subcontracting researchers were to use statistical methods capable of detecting small changes in measured variables and well-established techniques, that biologic end points must be sufficiently understood to ensure confident interpretation, that blind scoring must be used in laboratory analysis, and that field studies should produce only minimal impacts. The RFP discussed criteria for experimental design, including use of paired plots (treatment and control), exposure criteria for establishing treatment plots, requirements for characterizing ELF exposures at control and treatment plots, and such ambient monitoring as recording of climatic data. Annual reports and a yearly symposium that was open to the public would be required and would be used to review the progress of each study.

A preconstruction database was stipulated by the original Navy proposal and by the RFP. Information obtained from a monitoring program in the vicinity of the planned Michigan facility was judged to be sufficiently applicable to the Wisconsin facility to obviate further monitoring there. The RFP stipulated that the preconstruction database include data already available from federal, state, commercial, and private sources and data from several years of new field studies.

About 120 proposals were submitted in response to the 1982 RFP and were reviewed for merit by an independent panel of scientists; 11 were selected for support: wetlands, slime mold, Wisconsin birds, Michigan birds, small vertebrates, litter decomposition and microflora, upland flora, aquatic ecosystems, pollinating insects, soil arthropods and earthworms, and soil amebas. This group of studies included most of the specific flora and fauna listed in the Navy's original monitoring plan. The studies included collection of data on physiologic, developmental, behavioral, and ecological measures from dominant biota in upland, wetland, and riverine habitats near the Wisconsin and Michigan facilities. IITRI established a group of scientists to review the annual reports of the selected projects and attend the yearly symposia to discuss the status of the studies.

The studies began in 1982 to 1984. Observations were made before construction (Michigan only), during intermittent use, and during full operation of the facilities. Data collection for studies near the Wisconsin facility was completed in 1989. Data collection for studies near the Michigan facility was completed in 1993. The final reports of studies near both facilities were completed and peer reviewed by March 1996. The overall conclusion reported by IITRI is as follows:

> Although some investigators believe that a few biological changes may have occurred, all acknowledge that there were no consistent, unequivocal effects of ELF System operation on any of the variables they monitored. All conclude that the implications of their results are not indicative of adverse ecological significance (Zapotosky et al. 1996).

In 1994, the Navy, believing it essential to have an additional external evaluation of the ELF ecological monitoring program's activities and findings, requested that the National Research Council conduct an independent review and evaluation of the program.

THE COMMITTEE'S CHARGE AND APPROACH

The Committee to Evaluate the U.S. Navy's Extremely Low Frequency Communications System Ecological Monitoring Program was formed by the

National Research Council in 1995 to evaluate independently the objectives and design of the monitoring program, data-collection methods, data analysis, and interpretations. Important evaluation considerations included the following questions:

- To what extent did the physical measurements represent characteristics of EMFs appropriate for assessing possible ecological effects?
- Were proper experimental controls used to enable the detection of actual differences between exposed and unexposed organisms? Were known sources of concomitant variation controlled so as not to obscure a real effect? Were the experimental end points properly chosen?
- Were sample sizes adequate to reveal small effects?
- Was the sensitivity of experimental design, sampling, and analysis adequate to ensure a reasonable probability that effects, if any, would be detected?
- Were the experimental and observational techniques, methods, and conditions objective?
- Was the study internally consistent with respect to the effects of interest?
- Are the conclusions reached by the investigators well reasoned and well supported by the data?
- Are the results able to be confirmed by other investigators?

The ecological monitoring program and the committee's review addressed the possible effects of ELF-EMF exposures resulting specifically from the Navy's ELF communications system. The monitoring program and the committee's review did not address effects of ELF-EMF exposures in general (that is, from all sources). The committee did not analyze the data systematically. In some cases, the committee selectively examined the data in detail and in so doing uncovered inconsistencies in some of the analyses. The committee did not determine before its assessment whether any types of effects would be likely or unlikely, did not attempt to assess possible effects of ELF-EMF exposures beyond the study sites, and did not attempt to determine the possible effects of variations over time in factors, other than the operation of the ELF communications system, that might have affected the observations of the ecological monitoring program.

The committee reviewed a large number of relevant documents, including the original request for proposals, the original proposals, followup proposals, annual progress reports, reviewer comments, engineering reports, and the final reports produced for each of the 11 ecological studies. It found that often the written word did not convey all the analytic thinking and interpretation of the investigators, so on several occasions it discussed findings with the investigators for clarification.

The committee examined the responsiveness of study designs to program objectives, selection of study species and justification, selection of response variables and justification, the soundness of experiment designs and implementation from ecological and biophysical perspectives, statistical methods used for analysis, exposure assessment, presentation of results, and the validity and uncertainty of conclusions reached by researchers. This report presents the results of the committee's evaluations (biographic information on the members of the committee is presented in Appendix A). Although some data from the monitoring program are presented in this report, it is not intended to provide a comprehensive summary of the large amount of data contained in the numerous reports from the ecological monitoring program. Documents from the monitoring program are available to the public from the National Technical Information Service in Springfield, Virginia.

SPECIFIC THEORIES OF BIOLOGIC EFFECTS OF EMF EXPOSURE

From the beginning of its review, the committee was fully aware of scientific controversies surrounding the question of biologic responses to exposure to low-amplitude alternating EMFs in the ELF range. Scientists who have studied this question since the 1977 National Research Council report (NRC 1977) have offered a wide array of opinions (for example, see Wilson et al. 1989; NRPB 1992; ORAU 1992, 1993; NRC 1997). Some scientists believe that biologic responses attributed to EMF exposures have been reproducibly demonstrated; others are skeptical about the documentation of such responses; and yet others have expressed the opinion that such responses violate fundamental laws of physics and are therefore physically and biologically impossible. It was not within the scope of this committee's work to resolve such controversies, nor did it attempt to do so. An investigation of the extent to which controversial responses to low-frequency EMF exposure might present a hazard to human health was the mandate of the National Research Council's Committee on Possible Effects of Electromagnetic Fields on Biologic Systems (NRC 1997).

The present committee decided at the outset that in carrying out its mandate, it would be counterproductive to make assumptions about the possibility or impossibility of biologic responses to ELF EMFs. For example, if the committee had begun its work with a notion that testing data for consistency with Theory A of magnetic-field-induced biologic responses was most important, it might not have adequately considered data that were not consistent with Theory A. Instead, the committee evaluated all the Navy's experiments in each project according to criteria that are standard in all branches of scientific endeavor. According to the committee's charge, the experiments were reviewed for the coherence of the hypotheses being tested, the adequacy of the experimental design to test those hypotheses (including the adequacy of biologic species chosen, methods, statistical analysis, and statistical power), and the soundness of the conclusions that the investigators drew from their observations. The projects were not evaluated from the perspective of how a demonstrated biologic response to ELF electric-field or magnetic-field exposure might fit any particular opinion or theory about the plausibility of such responses. Where biologic responses to ELF EMFs were identified, physical mechanisms of field interaction with biologic molecules and biologic processes that might be affected by EMF exposure would necessarily be the subjects of future experimental investigations.

The ecological monitoring program researchers who conducted the individual studies have presented their results in scientific meetings, and some have been published in the peer-reviewed literature. Voluminous annual reports containing more detailed data are in the public domain. Members of the scientific community who wish to examine the results of individual experiments are encouraged to do so.

SCOPE OF THE REPORT

Chapter 2 of this report contains introductory material on exposure assessment, including engineering information on IITRI's measurements and general information on the concept of dosimetry. Chapter 3 contains evaluations of final reports of the 11 individual ecological studies performed for the Navy. The committee combined its discussion of studies of Wisconsin birds and Michigan birds. Chapter 4 discusses common issues that pertained to all the studies. Chapter 5 summarizes the evaluations and presents the major conclusions. It also recommends how the Navy might proceed with the information that it has received from the monitoring program and on how others who attempt a similar grand-scale monitoring program might organize its overall research and monitoring aspects.

2

EMF Measurements, Exposure Criteria, and Dosimetry

STUDYING THE EFFECTS of electric and magnetic fields (EMFs) on organisms involves accurate assessment of exposure to these fields and of the dose that an organism receives as the result of exposure. Exposure is a measure of the field strength of an electric or magnetic field immediately *outside* an organism over a specific period. Dose is a measure of the induced field strength *within* an organism over a specific period. The first section of this chapter describes how IITRI characterized the EMFs in the vicinity of the transmitting facilities. Later sections discuss the problems inherent in estimating doses and accounting for possible effects due to signal modulation.

The term "EMF" applies to an alternating field generated by moving charged particles. EMFs are characterized by their wavelength (expressed in meters) and their frequency (expressed in hertz). The wavelength of a field multiplied by its frequency equals the velocity of propagation. The full range of frequencies of EMFs is described as the electromagnetic spectrum. The "extremely-low-frequency" (ELF) designation is generally reserved for frequencies that range from 3 Hz to 300 Hz. Most equipment used for the generation, transmission, and distribution of electric power in the United States generates EMFs with a frequency of 60 Hz. The Navy's ELF Communications System uses a frequency-modulation principle called minimum-shift keying. In this type of modulation, the frequency is shifted between 72 Hz and 80 Hz (with a center of 76 Hz) depending on whether a code of "one" or "zero" is to be transmitted to a submarine (Zapotosky et al. 1996). The intensities of the electric fields are expressed in volts per meter (V/m), and magnetic fields are expressed in milligauss (mG). Additional information is provided by NIOSH, NIEHS, and DOE (1996).

CHARACTERIZATION OF ELECTRIC AND MAGNETIC FIELDS

To characterize the electric and magnetic fields near the ecological monitoring sites, IITRI measured the spatial and temporal characteristics of the following fields:

- A magnetic field in the air and the earth generated by the electric current in the antenna and ground terminals.
- An electric field in the earth that is the sum of the fields induced by the magnetic field and the current flowing from the buried ground terminals.
- An electric field in the air resulting from the difference in electric potential between the antennas and the earth or created as a byproduct of the electric field in the earth.
- The earth's static geomagnetic field.

IITRI provided the following dimensions of the EMFs near the transmitting facilities (see Table 2-1 for instruments used):

1. The ambient 60-Hz resultant[1] EMFs above the earth.
2. The unmodulated 76-Hz resultant EMFs above the earth.
3. The modulated 76-Hz resultant EMFs above the earth.
4. The root-mean-square (rms) values of harmonics of the 60-Hz and 76-Hz EMFs above the earth.

---

[1]"Resultant" is defined in the following way. The root-mean-square (rms) magnitudes of three rectangular components of the field are determined by either measurement or calculation. For fields that vary sinusoidally in time, the rms magnitude of each component is the zero-to-peak magnitude divided by the square root of 2. The resultant is the square root of the sum of the squares of those three rms values.

5. The earth potential differences in two orthogonal directions and, on the basis of this, the resulting electric field in the earth.

The ability to measure low-level magnetic fields depends on, among other things, the sensitivity of the instrument used to analyze the magnetic-field probe voltage. According to IITRI, the magnetic-field measurement equipment was calibrated by measuring the magnetic-field probe voltage output when placed in a 100-mG magnetic field (Haradem et al. 1994). It claims that this calibration is valid at smaller field levels on the basis of the fact that the probe, constructed solely of passive components, is known to have an output linear with respect to field intensity. The smallest full-scale sensitivity of IITRI's instrument (a Hewlett-Packard 3581A) was 0.1 μV, which corresponds to a magnetic-field level of about 0.2 mG. Measurements much lower than that (about 0.02 mG or lower) are not expected to be very accurate. Fortunately, the most-important reported levels used in establishing treatment and control sites and in analyzing ecological data were well above these levels and are expected to be accurate representations of the magnetic fields. However, reported field levels like 0.0002 mG are not expected to be accurate.

To eliminate the possibility of contamination of the ecological studies by harmonics or interactions between the power-line frequencies and ELF-antenna frequencies, a spectrum, field strength versus frequency, was measured. All the unwanted signals were found to be at least 30 dB below the level of the ELF-antenna frequencies and were therefore judged not to be contaminating the ecological studies. The rms values of harmonics of the 60-Hz and 76-Hz electric and magnetic fields above the earth were reported to be either below detection levels or "so low as to not be considered a confounder." Spectra measured by IITRI at the antenna terminals with the transmitter off and with the transmitter on yield data on the ambient 60-Hz fields and the 76-Hz fields (J.R. Gauger, IITRI, letter to U.S. Navy's Communications Systems Project Office, December 23, 1985). Although the spectral data reported represent a single day's observation (December 11, 1985), they confirm the statement quoted above.[2] Indications of the natural Schumann resonances in the earth's

---

[2]The power-line 60-Hz signal is 30 dB below the transmitter 76-Hz signal; and the strongest harmonic of the power-line frequency (at 300 Hz) is at least 60 dB below the transmitter signal. The spectra show harmonics of the ambient power-line frequency out to the 17th harmonic and of the transmitter frequency out to the 11th harmonic. The harmonics of the transmitter are down from the fundamental by 35 dB at 216 Hz and 240 Hz, by 50 dB at 144 Hz and 160 Hz, and by 55 dB at 360 Hz and 400 Hz.

atmosphere and the measured behavior of the harmonics also lend credence to the quality of the observations. A few obvious errors in the labels on the spectra are consistent with the problems of field observations and were easily removed. However, such errors were relatively small.

TABLE 2-1 Instruments Used by IITRI to Measure EMFs

An IITRI-constructed single-axis magnetic-field probe with flat frequency response. The output of the sensor was a voltage proportional to the magnetic field.

A single-axis electric-field probe that consisted of a spherical sensor with an optical-fiber link to a receiver. The output of this probe was a voltage proportional to the electric field.

A probe to measure the electric field in the earth that consisted of two orthogonal electrode pairs. The output was the voltage difference between the electrodes in a pair.

A Hewlett-Packard 3581A signal-wave analyzer. This device, a frequency-selective rms-calibrated voltmeter with an adjustable bandwidth, was used to measure the voltage output of the three probes mentioned above.

An IITRI-constructed 60-Hz notch filter to filter out the 60-Hz contribution from the modulated 76-Hz measurements.

A Walker Scientific FGM-3D1 fluxgate magnetometer for measuring the static magnetic field in the earth.

An electric-field measurements EMDEX II magnetic-field meter for long-term measurements of the resultant magnetic field in the frequency range 40-400 Hz.

An IITRI-constructed instrument for measuring the electric field in the earth as a function of depth.

EXPOSURE CRITERIA FOR SITE SELECTION

Any source of electric and magnetic fields (such as the ELF transmitting facility and antennas) creates fields essentially everywhere. As one moves farther from a source, the field intensities become lower, either because of distance or because of attenuation due to obstacles (in the case of electric fields). However, one will then be moving toward other sources, and the EMFs that they generate will increase in intensity as one moves closer. It is therefore not possible to select a control site where there is no exposure to ELF EMFs generated by the Navy's transmitting facility and antennas. It is possible only to select sites that have different levels of exposure to the EMFs generated by the antennas and to those generated by other sources, such as power lines.

IITRI helped researchers to select study sites for the ecological monitoring program by determining whether they were in the treatment or control category. The specific criteria used to determine whether a site was treatment or control were as follows:

$$T(76 \text{ Hz}) / C(76 \text{ Hz}) > 10$$

$$T(76 \text{ Hz}) / T(60 \text{ Hz}) > 10$$

$$T(76 \text{ Hz}) / C(60 \text{ Hz}) > 10$$

$$0.1 < T(60 \text{ Hz}) / C(60 \text{ Hz}) < 10$$

where T(76 Hz) is the treatment-site exposure due to the ELF communications system, T(60 Hz) is the treatment-site exposure due to power lines, C(76 Hz) is the control-site exposure due to the ELF communications system, and C(60 Hz) is the control-site exposure due to power lines.

In other words, the intensities of the 76-Hz EMFs at a treatment site had to be 10 times as large as the intensities of the 76-Hz EMFs at the control site. In addition, at both treatment and control sites, the intensities of the 76-Hz EMFs due to the antennas had to be 10 times as large as the intensities of the 60-Hz EMFs due to nearby power lines. Finally, the ratio of the intensities of the 60-Hz fields at a treatment and control site had to be between 0.1 and 10. Those criteria were applied to EMFs in the air and in the earth during full-power operation of the relevant transmitter antenna. However, there was no evidence to indicate, a priori, that a one-tenth reduction in exposure should lead to a reduction of one-tenth (or less) in the effect that might be observed with full exposure.

Variation of field intensities with distance means that each site is exposed to a spatial gradient of intensity, rather than a uniform intensity across the site. Sites were therefore classified according to annual measurements taken at the same point each year.

To isolate the effects of the ELF EMFs, paired treatment and control sites were intended to be as alike as possible with respect to ecological variables, including soils, foliage, species abundance, and temperature, depending on the focus of the study. For example, the wetlands study required similar bogs for treatment and control sites, whereas the pollinating insects study required sites with similar flower abundances. Several study teams had difficulty in identifying pairs of sites that met both the exposure and ecological criteria, as discussed in Chapter 3.

EXPOSURE DATA SUPPLIED TO RESEARCHERS

IITRI made available to researchers data on magnetic fields and electric fields in the air, and electric fields in the earth. IITRI provided extensive data on specific measurements of ELF EMFs at each site to the ecological monitoring teams. The purpose of these measurements was to allow the monitoring teams to determine indicators of exposure for different parts of the treatment sites and to attempt to relate indicators to appropriate measures of ecological effect.

IITRI also made data about transmitter on and off times available to the ecological monitoring teams. These data could be used to determine whether a site characterized as a treatment site was actually exposed to ELF EMFs from the antenna at any particular time. That is important because the transmitter was not on continuously and a treatment site is exposed only when the transmitter is on.

USING FORMULAS FOR PREDICTING ELECTRIC AND MAGNETIC FIELDS

Electric and magnetic fields can be characterized either by physical measurement or theoretically, although there are difficulties inherent in both methods. Physical measurement of field strength can be difficult because of the limitations of measuring equipment and the gradients in field strength caused by distance from the source and alterations in terrain. Such fields are often too complex to be adequately characterized using simple formulas. The 76-Hz magnetic field was relatively steady and could have been well characterized spatially with simple formulas. However, physical measurements were necessary to characterize the electric fields. The spatial dependence of the electric fields in the earth could not be predicted with simple formulas because of spatial variability in the earth's conductivity. Furthermore, there was a (usually modest) temporal variation due to daily and annual changes in the earth's conductivity. It was necessary to do careful surveys of the spatial and temporal variation of electric fields in the earth. The electric field in the air has characteristics that fall between those of the magnetic field and the electric field in the earth. In an open area with level terrain, this field is well characterized with simple formulas. However, in the presence of obstacles, such as trees, it is difficult to calculate electric fields in the air, and measurements provide only snapshots because the field is varying. For example, wind moving through foliage will cause variability in the electric fields in air.

IITRI provided several formulas that can be used to calculate the ELF EMFs in the vicinity of the antennas. Not all the limitations of these formulas have been stated in the reports received by the committee. The authors have correctly pointed out that the formulas are only appropriate for field points on the earth's surface and close to the antennas. However, there are other restrictions. First, the formulas are all restricted to the quasistatic frequency range, so all relevant distances should be substantially smaller than a wavelength in the earth. Second, each assumes an ideal flat homogeneous earth. Some of these assumptions are valid, others are not and would preclude use of the formulas. For example, it is reasonable to calculate the magnetic fields (either in the air or in the earth) with the simple Biot-Savart law provided by IITRI if the antenna current is known (which it is). But it is usually not valid to assume that the earth conductivity is either homogeneous or independent of time for calculations of electric fields in the earth. If it is necessary to account for variations in conductivity with depth or horizontal position, the formula given by IITRI is not valid. In fact, it was found that the conductivity varied substantially over some of the study plots and that it depended upon environmental conditions and hence time of year. That is why one should be very careful in using that formula and why more measurements of electric fields in the earth were needed. The care required in using the formula for electric fields in air is between these extremes. For uniform flat terrain with no obstructions, the formulas provided are satisfactory (provided that antenna voltage is known). However, in a grove of trees, there is much distortion of the fields, and the formula provided by IITRI would be of no use.

Of the formulas provided by IITRI, only the magnetic-field formula was widely used by researchers. This formula was used only in efforts to interpolate the magnetic fields at points within study sites. For the upland-flora study, measurements of the 76-Hz magnetic field were made at several points near the Wisconsin antenna. Table 2-2 compares measured with calculated fields for an antenna height of 13.7 m and an antenna current of 150 A. The measured and calculated values agree reasonably well.

Unfortunately, no measurements of antenna-to-ground voltage that could be used to validate reported measurements of electric fields in air were reported by IITRI. However, this voltage could be estimated in the following way. According to Dill (1984), one design criterion for the ground terminals was achievement of a maximal total ground resistance of 6 ohms for both grounds. On the basis of that number and the assumption that the antenna resistance is much smaller than the ground resistance, the antenna-to-ground voltage at full current (150 A) is 900 V. According to the formula provided by IITRI, the electric field in the air directly below the antenna would be about 25 V/m. The measured electric field under the antenna for the small-mammal and nesting-bird studies was 10-40 V/m. It can be inferred from this result and the theoretical estimates that the measurements are in reasonable agreement.

DOSIMETRY

As stated above, the EMF quantities relevant to ELF interaction with biologic systems are the exposure (field strength immediately *outside* the organism over a period of time) and the dose (induced field *inside* organisms over a period of time). The latter quantities can be expressed in terms of induced electric-field strength, magnetic-field strength and induced current or current density. Dosimetry involves assessment of the magnitude and distribution of induced fields and currents within biologic organisms that are exposed to ELF EMFs. The induced fields and currents not only are functions of the externally imposed EMFs, but are determined by the EMF properties and geometry of the exposed organism and any nearby objects. Such induced fields and currents at ELF should not be expected to constitute the dose.

TABLE 2-2 Comparison of Measured and Calculated Magnetic Fields

Site	Horizontal Distance to Antenna, m	1993 Measured Magnetic Field, mG	Calculated Magnetic Field, mG
4T2-13	28	10.3	9.7
4T2-12	58	5.5	5.0

Dosimetric measures of the induced fields and currents were not part of the study design. However, to provide some indications of relative strength of induced electric fields in various biota exposed to 76-Hz EMFs, the committee has performed some analyses with simple models to serve as an index to the induced fields within an organism. The ELF-EMF exposure environment was characterized at control and treatment sites through periodic surveys. This environment included 76-Hz fields produced by the ELF communications system, 60-Hz fields from power lines, and the earth's magnetic field. Because the wavelength at 76 Hz is much longer than the longest dimension of the organism, the quasistatic-field theory can be appropriately applied to calculate the induced electric field inside the body of the organism (Michaelson and Lin 1987).

Briefly, the calculated results suggest that induced electric fields in insects, birds, and small vertebrates are fairly low for exposures to external electric fields up to 5,000 mV/m and magnetic fields up to 50 mG. In contrast, electric fields induced in hardwood stands by the same EMFs could be substantial. Calculations based on these simple models suggest that the field induced by a vertically oriented electric field in a 25-m tree could be as high as 5,000 mV/m and that induced by a horizontally oriented magnetic field as high as 29.8 mV/m. The applied or impinging electric field would decrease in strength with distance from the antenna wire and because of shielding. However, magnetic-field strength would be attenuated away from the antenna wire only by distance. Therefore, at greater distances from the antenna, the field induced in tree stands by a horizontal magnetic field could become a dominant factor in the resulting dose. (See Appendix B for more detailed information.)

Researchers involved in the ecological monitoring program were not asked to estimate doses received by biota from the Navy's ELF transmitting antennas. Only the researchers involved in the upland flora study attempted to do so. Because sufficient information was not available on dosimetry, the committee determined that it was not possible to extrapolate the monitoring program's findings to other situations that might be comparable with exposure conditions.

DIFFERENCES IN EFFECT BETWEEN UNMODULATED 60-Hz AND MODULATED 76-Hz SIGNALS

As mentioned previously, the ELF electric and magnetic fields generated by the communications system antennas are frequency modulated between 72 and 80 Hz (with a dominant frequency of 76 Hz), unlike power-line EMFs, which are unmodulated at 60 Hz. Unfortunately, there is little information on the differences between the effects of modulated and unmodulated frequencies. Most of the research undertaken over the last 25 years to understand the biologic effects of low-frequency EMFs has focused on exposure to unmodulated EMFs at power-line frequencies of 50-60 Hz (see, for example, Anderson 1990; ORAU 1992; Tenforde 1996; OTA 1989; NRC 1997). There has been little research on the effects of the modulated 76-Hz signals produced by the ELF communications system.

CONCLUSIONS REGARDING EMF MEASUREMENTS

IITRI has done a good overall job of characterizing the ELF electric and magnetic fields in the vicinity of the treatment and control sites. In cases in which it became obvious that more information was needed, IITRI was responsive and performed additional measurements. Specific conclusions are as follows:

- Although there were some minor questions about instrument design, it appears that the associated errors were small and would lead to no changes in IITRI's conclusions about the measured data.
- The spatial and temporal variation of the magnetic fields above the earth near the treatment and control sites has been well characterized.
- The spatial and temporal variation of the electric fields above the earth in open areas near the treatment and control sites has been well charac-

terized. In shielded areas, such as near trees, more extensive measurements were necessary. When requested, these were provided.
• The electric fields in the earth depend on local earth conductivity, so they require a more-thorough measurement survey to characterize them. When requested, IITRI provided engineering support for these measurements.
• The electric fields in the earth have been examined in light of annual changes in earth electric characteristics. Most variations were modest, but there were daily and annual changes in these fields.

3

Evaluation of Final Reports of Individual Studies

INTRODUCTION

THE STRUCTURE OF THE ecological monitoring program was segmented, rather than integrated, in that the research teams worked independently of each other. The purpose of IITRI's solicitation in March 1982 was to attract subcontracting researchers to develop and conduct separate ecological monitoring studies that would determine whether low-level, long-term electric and magnetic fields (EMFs) and gradients produced by the ELF communications system would affect vegetation or wildlife in and near the system area or otherwise result in changes in individual organisms or their communities. Eventually physiologic, developmental, behavioral, and ecological aspects of predominant organisms in upland, riverine, and wetland habitats near the Navy's ELF transmitting facilities were monitored for possible effects of EMFs produced by the Navy's antennas. The organisms and ecological relationships selected for monitoring were chosen because they were judged to be important to their ecosystems and to be of interest to local residents (Zapotosky and Gauger 1993).

The monitoring program studies were designed to compare data collected at control sites with data collected at treatment sites. As discussed in Chapter 2, the paired sites were intended to have matched environmental factors but to be dissimilar in the magnitude of their exposure to the 76-Hz EMFs generated by the communications system antennas. Sites exposed to those 76-Hz EMFs were established by locating treatment sites near or within the rights-of-way for the antennas; control sites had to be far enough from the communications system that EMF intensities resulting from antennas would be substantially lower than those at treatment sites. However, control sites had to be close enough to have environmental factors similar to those of their matched treatment sites. Siting criteria called for intensities of the 76-Hz EMFs at treatment sites to be at least 10 times those at control sites, for intensities of the 76-Hz EMFs at treatment sites to be at least 10 times those of the 60-Hz EMFs at treatment sites and control sites, and for intensities of the 60-Hz EMFs at treatment sites to be within 10 times those at control sites (Haradem et al. 1994).

This chapter presents evaluations of the 11 final ecological reports on the following topics: wetlands, slime mold, Wisconsin birds, Michigan birds, small vertebrates, litter decomposition and microflora, upland flora, aquatic ecosystems, pollinating insects, soil arthropods and earthworms, and soil amebas. The committee combined its discussion of Wisconsin and Michigan birds in this chapter. The committee used the following criteria to evaluate the reports: adherence to the original project proposal, adequacy of selection of species and response variables, adequacy of experiment design and implementation (including biologic and ecological sampling techniques, physical measurements, and statistical power), responsiveness to reviewers' comments while studies were being conducted, presentation of results (including consideration of alternative analyses or hypotheses and interpretation), and appropriateness of conclusions (including validity and uncertainties). The committee found that the various criteria did not warrant the same amount of discussion for each study.

WETLANDS

Project Proposal

The authors of the wetlands final report (Guntenspergen et al. 1989) pointed out that wetlands in the upper Midwest are sensitive ecosystems and are common near the ELF communications system sites, especially the Wisconsin location. They pointed out that past studies on effects of ELF EMFs indicated that plant membranes might be affected by electromagnetic radiation.

Therefore, they proposed to look for changes in plant competitive ability through measurement of plant or ecosystem functions that were related to membrane functions. Possible changes included leaf diffusion resistance, foliar nutrient content, changes in functions related to transport of water and nutrients across membranes, and decomposition by microorganisms, all of which depend heavily on secretion and adsorption through membranes. The basic null hypothesis of the wetland-monitoring project was that ELF EMFs resulting from the operation of the Navy antenna have no effect on selected ecosystem variables.

SYSTEM, SITE, AND SPECIES SELECTION

Much of the initial, pilot-study year of this project was spent in establishing study sites and testing methods. Stearns et al (1982) described five northern wetland vegetation types in Wisconsin: northern conifer swamp, shrub wetland, emergent marsh, northern sedge meadow, and open bog. Because the northern conifer swamp was common near the ELF location in Wisconsin and offered all life forms—including trees, shrubs, herbs and nonvascular plants—it was chosen as the ecosystem type for monitoring.

The authors of the proposal recognized the heterogeneity of the region near the ELF antenna, as well as the heterogeneity among stands of the same type of wetland. That created a problem in site selection but was addressed through identification of sites with close similarities in vegetational composition and environmental characteristics. The former was determined through use of contingency tables and similarity indexes (such as Sorensen's Index), and the latter through measurement of soil-water temperature and pH, cation concentrations, and redox potential. In general, it was thought that sites with similar vegetation would have similar, but certainly not identical, environments. More than 200 potential sites were selected from aerial photographs; these were reduced to 50 sites for priority-setting, and then 15 sites were selected for measurement of 60- and 76-Hz EMFs in potential study plots, one per site. Eleven sites were eventually chosen for initial summer (1983) studies.

Location of "similar" sites was based on relative EMF intensity. Electric-field strength measured by IITRI was used to establish intensity gradients; these were called background, intermediate, antenna, and ground to correspond to location of sites relative to ELF facilities. No true control could be established, because the ELF communications system was already functioning, so the background sites were considered as control sites. In most cases, three sites were used for each exposure scenario, but only two ground sites were used within one large peatland. Although the selection of study sites satisfied the criteria established by IITRI and a gradient of exposure from treatment sites to control sites existed, the three treatment sites varied among themselves in terms of EMF intensity. For example, 1987 field measurements showed that the plots within the antenna treatment sites ranged from 0.053 to 0.196 V/m in electric-field intensity and from 6.1 to 19.8 mG in magnetic-field intensity.

Selection of the northern conifer forest wetland allowed use of tree, shrub, and herb species, when appropriate, for determination of response variables associated with plants. Final experimental species were not selected until 1985. Labrador tea (*Ledum groenlandicum*) was selected as the primary species for measuring stomatal resistance after other species were tried; for instance, the leaf anatomy of black spruce (*Picea mariana*) made stomatal measurement difficult. Labrador tea is a common shrub throughout the conifer wetlands of Wisconsin. Labrador tea leaves replaced pure cellulose sheets in the decomposition studies for testing of "natural plant materials" and to improve within-site measurement consistency. Black spruce, Labrador tea, the shrub leather leaf (*Chamaedaphne calyculata*), and the herb false solomon's seal (*Smilacina trifolia*) were used for foliar nutrient content; this permitted comparisons across life forms. Nitrogen fixation studies, initially on alder and then on moss and peat, were dropped during the study period.

RESPONSE VARIABLES

Response variables chosen for the wetland studies and used throughout the study period all were related to membrane-associated functions. The choice was based on a National Research Council (NRC 1977) report and other studies on effects of EMFs that indicated that biologic and ecological responses to EMFs were most likely in functions associated with membranes. Stomatal resistance was chosen as a response variable because it is associated with water transport across membranes and all vascular plants could potentially be influenced through this leaf function. Nutrient content of leaves was examined because it is closely related to nutrient transport across membranes in root cells. Decomposition was examined because it is associated with across-membrane secretion of enzymes by microorganisms and adsorption of decomposed cellulose and other organic compounds. Other processes could have been examined, such as growth rates or species composition changes, but they were not, because they were considered likely to result in too much variability within sites and were considered to be long-term response variables. Nitrogen fixation was examined for a couple of years, but that effort was eventually dropped from the program because the method was not reliable.

Experiment Design

The wetland monitoring study was conducted from 1983 through 1987. The first year was used for selecting sites, developing and evaluating protocol, and beginning preliminary sampling. After experimentation with a rectangular plot design in the first year in which all the experimental measurements were made in a regular array throughout the plot, the study team settled on a 70 x 15-m rectangular plot for all experiments, oriented with the long axis parallel to the closest antenna. Six square subplots were designated in the rectangular plot with centers (where shallow groundwater wells were placed) 10 m apart. All measurements of response variables were taken within these subplots; all selected species were within each subplot.

Environmental data were collected monthly from May to September (the frostfree period). ELF EMFs were measured once a year and assumed to be constant over the year. The antennas were capable of operating at full strength from 1985 through 1987, when most of the established protocols were in place; however, the antennas were not on full-time during this period, and it is not certain to what extent the off periods were taken into account in the study. That might be important for response variables that respond instantaneously or in a short term.

Biologic Sampling

The final report presented three biologic measurements used to determine possible effects of ELF EMFs on wetland ecosystems in the vicinity of the Wisconsin transmitter facility (Guntenspergen et al. 1989): stomatal resistance, foliar nutrients, and decomposition.

Stomatal Resistance   The primary results of stomatal-resistance tests were based on responses of Labrador tea, a common species in every study bog. Other species were tested both because of ease of measurement and because of response to different light levels. Spruce and smilacina were dropped because of measurement difficulties. The equipment used for measurement of stomatal resistance was a null-balance diffusive-resistance porometer, which measures the rate of water-vapor diffusion through the stomata. It is standard equipment for such measurement, and the method is easily replicated if needed.

Stomatal resistance was measured in leather leaf and Labrador tea under different light intensities. Labrador tea, least responsive to differences in light especially at low intensities, was selected as the species for testing the response of plant stomatal resistance to ELF EMFs. That choice essentially eliminated light as one of the independent environmental variables that might have to be considered a covariate in later statistical analyses. Selection of only one species, however, implied the assumption that all other species would respond to ELF EMFs in a similar fashion. If different species had different stomatal-resistance responses to light levels, might this indicate different responses to other external variables? That is partially addressed by the results, which include data on leather leaf, which was dropped as a consequence of tests on light intensities.

Measurements were made during four periods at all 11 study sites—in August and September 1986 and 1987. Measurements were made over several days in each sample period, and there was an attempt to stratify measurements to cover all environmental variables, especially light intensity and cloud cover, for the background, intermediate, antenna, and ground sites. The number of samples taken was doubled between 1986 and 1987 to allow for resolving 20% differences in means at $p = 0.05$ with an 80% probability. The increase in samples increased variability in the measurements of external environmental conditions. That situation is commonly found in field sampling. An increase in sampling frequency might reduce the statistical significance of the results because it increases variability in sample measurements.

Foliar Nutrients   Changes in foliar nutrient concentrations in plants growing in a relatively nutrient-poor environment were considered a possible indicator of the condition of various plant biochemical pathways. The initial species for foliar-nutrient sampling were a shrub (leather leaf), an herb (smilacina), two sedges, and a tree (black spruce). Labrador tea was substituted for the sedges because destructive harvesting of these sedge species might have damaged their limited populations. Sampling periods were based on phenology; thus, herbs were sampled earlier in the season than shrubs. Sample size was increased several times over the early years of the study to improve the power of the statistical analysis, eventually reaching 396 (6 × 6 × 11) samples per species per sample date. Only current-year foliar tissue was collected. Standard methods were used for preparation and analysis for calcium, magnesium, and potassium. Samples from 1987 were also analyzed for manganese and phosphorus. For quality control, National Bureau of Standards (NBS) standards were analyzed with field samples, and spikes of known-cation standard solutions were added to field and NBS samples.

Decomposition   Two approaches were used for decomposition studies. The first used pure cellulose as a substrate, and the second used Labrador tea leaves. The use of pure cellulose was intended to provide a uniform substrate for decomposition. If preweighed samples of cellulose were placed in a fiberglass bag and inserted vertically into the peat, less variation was expected among all samples. However, the cellulose became soft, adhered to the bag, and could not be retrieved fully for posttreatment weighing. Labrador tea leaves were used instead of cellulose. Bags with about the same amount of leaf material were mixed and randomly selected for placement in the bogs. The use of Labrador tea leaves resulted in less within-group variance.

The decomposition bags were allowed several months for incubation in situ. The bags were retrieved, and the cellulose or leaf samples were removed, cleaned of foreign material, dried, and weighed. Weight loss indicated decomposition. The duration of incubation varied from 4 to 12 months.

Environmental Characteristics

The primary environmental characteristics measured during the study, excluding ELF-EMF exposure levels, were those of the interstitial water within each wetland site. The characteristics of the bog water were considered to influence decomposition rates and root activity. Shallow groundwater wells were placed in the center of each of the six subplots in the study plot at each site. Water-quality measures were depth to water table, depth to anaerobic zone, reduction-oxidation potential, specific conductance, temperature, pH, and calcium, magnesium, and potassium contents. Dissolved organic carbon was measured in 1984 only. Water samples were prepared with standard methods.

No data in the wetlands report indicate regular measurement of ambient temperature, rainfall, or other external climatologic conditions. Measurements of water quality, light intensity and leaf temperature made during various sampling periods appear to have been considered sufficient for evaluating the influence of the external conditions.

Statistical Methods

Statistical methods were chosen to test the null hypothesis that ELF EMFs resulting from the operation of the Navy antenna have no effect on selected ecosystem variables. Researchers primarily used nested analysis-of-variance (ANOVA) models to examine treatment and control groups. Stepwise-multiple-regression models were used to explain the variance in the dependent biologic variables (stomatal resistance, foliar nutrients, and decomposition rates) on the basis of environmental variables (such as water-quality data) and ELF-EMF data. Significance levels ($p=0.05$) of the two models were compared. In a few instances, the models did not agree, and lack of significance within one model was selected as the appropriate test of the response variable (i.e., it was not significantly influenced by ELF EMFs). Reviewers of the project over the period of the study sometimes questioned the use of stepwise multiple regression because of the interdependence of the variables, even those considered independent. In ANOVA, the variables were usually considered covariates; in multiple regression, they were treated as independent. Treating variables in this manner is a common practice in ecological studies, the understanding being that no variables in an ecosystem are truly independent. The use of stepwise regression attempts to alleviate this concern.

To show the relationship of environmental variables to decomposition rates, all environmental data collected during the incubation period were subjected to principal-components analysis (PCA), an approach that reduces the number of independent variables to a few "composite" variables. The principal components representing environmental data were then regressed (in stepwise fashion) against decomposition rates.

Several times during the study period, the number of samples was increased to increase the power of the statistical analysis. That was done mostly for the foliar nutrient analysis. Other levels of sampling were considered sufficient, after preliminary studies, to achieve the established level of confidence at a 0.05 significance level.

Quality Assurance and Quality Control

Quality assurance and quality control, especially as related to chemical analysis, were addressed in the methodology discussion of the final report. Methods selected for stomatal resistance were standard and repeatable. Decomposition-rate methods were straightforward. The research design established for this study helped to avoid pseudoreplication and allowed sound ANOVA of data from study sites.

Exposure Assessment

IITRI provided ELF-EMF exposure data at the background, intermediate, antenna, and ground sites. IITRI measured ELF EMFs annually at each sample plot at the study locations. The 1987 annual report (Guntenspergen et al. 1988) made it obvious that the investigators took into account the spatial EMF gradient resulting from antenna operation and used it in regression analyses.

The ELF communications system was not operated continuously; study sites were therefore not consistently exposed to ELF EMFs. Different portions of the antennas were turned on and off several times each day, with varying modulations, frequencies, current intensities, and phase angles. The Navy provided IITRI and researchers with detailed logs of antenna activity. The final wetlands report does not indicate whether the antennas were on or off during field measurements or whether information on the antenna operation was used in data analyses.

Response variables would likely have varied in their sensitivity to antenna operations. For example, annual decomposition rates could be related to the annual EMF measurements, but stomatal resistance might have an immediate cellular response to external factors, such as light, temperature, and ELF EMFs. Use of short-term response variables, such as stomatal resistance, could have created analytic difficulties. Field measurements of these very short-term response variables would have had to be timed in coordination with antenna operations to ensure measurement of possible EMF influences that might not be observable when the antenna is off.

In 1987, stomatal resistance was measured in August, the same month that IITRI measured exposure levels at each study plot. There is some evidence in the 1987 annual report (p. 17) that the researchers examined the on-off status of the Wisconsin transmitting facility relative to their field sampling.

If these measurements occurred at the same time, the exposure assessment of this study took into account the need to measure short-term responses while the antenna was actually on. The final report does not address the coordination of field measurements with antenna activity. If the measurements were not coordinated, the use of a short-term response variable, such as stomatal resistance, is questionable.

RESPONSE TO REVIEW

The researchers received comments from reviewers starting with the 1983 annual report (Stearns et al. 1984). Common comments were related to site characterization, use of specific response variables and timing of measurements, and level of sampling for adequate statistical analysis. The study design was modified to accommodate some of the critics' comments; others were addressed through explanations of why changes were not made.

For example, a criticism of site selection was lack of extensive soil data. The researchers characterized similar sites as those with similar peat substrates, but then emphasized measurement of interstitial water at each site as more appropriate for studying substrate composition.

The use of stomatal conductance as a response variable was acceptable to the reviewers, but they were concerned that because this variable is so closely influenced by light and temperature, it needed to be measured at the same time of day under similar conditions of light and temperature to yield useful comparisons. The researchers selected the species with the least response to light variation because it was impossible to make all measurements at the same time of day or even on the same day.

The use of foliar composition was questioned because it is not considered appropriate for determining soil nutrient availability in agricultural systems. The researchers pointed out, however, that foliar composition is commonly used in natural ecosystems as an indicator of plant condition and therefore was appropriate for this study. Although the final statistical approaches did not fulfill all the requirements of the reviewers, the study design was altered in some cases to increase sample size. Inappropriate uses of statistical terms, pointed out by reviewers, also were corrected in the final report.

In most cases, the investigators did respond to reviewers' comments, thus producing an improved final document.

Presentation of Results

Alternative Hypotheses

Results were presented in different forms: graphical presentations with bar graphs (often showing standard errors), tables with nested ANOVA analyses, and tables with stepwise-multiple-regression analyses. The initial proposal suggested that ANOVA would be the appropriate statistical approach to test the null hypothesis of no difference between or within treatment and control sites. Development of data sets also demonstrated to the researchers that there was a need to attempt to explain the variances of the biologic responses between and within treatments, in addition to assessing them. Although no alternative hypotheses were presented, use of stepwise multiple regression to assess the significance of independent environmental variables indicates possible consideration of an additional hypothesis that variances in dependent biologic responses are explained no more by natural external environmental factors than by ELF EMFs.

Interpretation

Interpretation of results was based primarily on comparison of the two statistical approaches to the empirical data. If the ANOVA models indicated that there was no greater difference between study sites than within study sites at the 0.05 significance level, the researchers interpreted that as conclusive. They checked their interpretation by applying the multiple-regression models that used environmental characteristics (sometimes considering biologic variables as independent, such as leaf nutrients for stomatal-resistance comparisons) and ELF-EMF characteristics (in most cases in a PCA form). In many cases, neither environmental nor ELF-EMF principal components accounted for much of the variance of the response variable. In a few cases, a significant correlation for a response variable was found using the regression model that was not found using the ANOVA model. The researchers interpreted the model showing no significant correlation to be correct. That raises the question of the level of confidence selected to demonstrate statistical significance of the ANOVA and multiple-regression models. For this study, a significance level of 0.05 was chosen. Considering the variability of the ecosystems being studied, this is probably appropriate, although many statisticians might consider it no more than an indication that the variance of the results depends on chance.

How Well Researchers' Conclusions Were Supported

The general conclusion of the researchers was that ELF EMFs generated by the communications system had no measurable effect on the biologic functions used as surrogates of wetland ecosystem responses. The functions studied were stomatal resistance, foliar nutrient content, and decomposition rates. All those functions are in some way influenced by across-membrane movement of water, nutrients, or enzymes. Using nested ANOVA to test the significance of variance in responses between and within treatments and the control, they concluded that the few occurrences of significance were at a level not much greater than would be due to chance (about 5% for this study). Using stepwise multiple regression to explain the variance of biologic responses, primarily to see whether ELF-EMF principal components might be among the most significant independent variables in the development of the coefficient of determination, they concluded that environmental variables explained more of the variance in the biologic-response variables than did ELF EMFs—and not much at that. There were very few statistically significant ANOVA tests, and the multiple-regression analyses of these same variables were not statistically significant. The researchers did not attribute biologic importance to the results from any one test that showed a statistically significant response to ELF EMFs; they noted that such results are expected because of chance alone when many similar tests are performed, as was the case in this study. Instead, they looked for statistically significant effects attributable to ELF EMFs that were repeated over months or years. This appears to be an appropriate approach because response variables were statistically significant in few cases and were always the results of only one test

Because the experiments were straightforward, essentially well designed, and properly analyzed, the conclusions were well supported by the data. Although none of the studies was simple, the methods were sound and reproducible and gave the researchers confidence in their conclusions.

COMMITTEE CRITIQUE

This study was properly designed from the beginning, although minor changes might be necessary to improve it. Some constraints were placed on the study because of prior antenna operations and complexity of the landscape. It also had built into its design the ability to modify the procedures when necessary to improve statistical power or to substitute a material that would produce more-consistent results. For example, the number of leaf samples taken for foliar nutrient content was increased several times during the first few years of the study, and substitution of Labrador tea leaves for pure cellulose represented not only use of a natural component of the ecosystem, in addition to a cellulose source, but also a material that gave more consistent responses.

Initial establishment of study sites was compromised in that the Navy's ELF facility was already in operation at the time of the first response measurements, so no true controls could be established. It was never explained why the background locations, which were intended to represent the low end of ELF-EMF exposure, had to be near the antenna. Was that prerequisite established to ensure that ambient conditions would be similar for all sites? Were there no northern conifer bogs at some distance from the ELF antenna with all the same characteristics as those selected as exposure sites? The researchers considered 200 sites on the basis of aerial photographic comparisons. How far from the antenna were some of these sites, and could some have been used as better "control" sites?

Researchers justified their choice of response variables on the basis that they represented biologic functions that were associated with across-membrane processes, membrane functions being known to respond to EMFs. Other ecosystem functions were not considered in the proposal or in continued development of the study over the first few years. A substudy on nitrogen fixation, also associated with membrane functions, was attempted but discontinued. For example, other growth processes might have been considered. In the decomposition study, results indicating statistically significant differences in rates of decomposition were apparently compromised by the differential growth of moss over the decomposition bags. Bags at sites with higher ELF-EMF intensities had almost total moss cover; those in the background were only partially covered. Although growth of moss was not an initial response variable, might not this phenomenon, albeit discovered well into the study, have triggered some controlled studies on growth of nonvascular plants in the bog? Because the moss-growth phenomenon was not controlled, there was no way for the study to address the potential relevance of this occurrence.

The use of a short-term response variable, stomatal resistance, might have created an analytic problem relative to exposure assessment. The on-off cycle of the antenna might have influenced the temporal response of this variable, but antenna operations do not appear to have been factored into the analysis or interpretation. However, the spatial variance of exposure levels within treatment and control sites was taken into account. If biologic field measurements were coordinated with ELF-EMF exposure measurements (this might have been possible in August 1987), then this study could have related specific exposures to short-term response variables.

In light of the basic soundness of the design and analysis, the results, interpretation, and conclusions of this study are in all likelihood appropriate and acceptable. They have demonstrated that, according to the responses of the variables monitored, wetlands, an ecosystem characterized by water through which ELF EMFs readily pass, appear not to have been significantly influenced by ELF EMFs.

SLIME MOLD

Project Proposal

The hypothesis tested in this program was that exposure to weak ELF EMFs generated by the Wisconsin antenna would induce observable physiologic alterations in *Physarum* (a type of slime mold). *Physarum* cells were directly exposed to ELF EMFs generated by the Wisconsin antenna and to a laboratory simulation of the ELF EMFs at the Wisconsin ground terminal site.

Species Selection

Physarum polycephalum was selected for two main reasons. First, the results of laboratory-based research had indicated that weak ELF EMFs, similar to what might be associated with the Navy ELF communications system, had effects on *Physarum*. Exposure to such fields had been reported to lengthen, reversibly, the cell cycle and to lower, reversibly, the respiration rate and the adenosine triphosphate (ATP) concentration of the cells. (Brayman et al. (1985) attempted to replicate the results of the original laboratory study but could not. The committee was not asked and did not attempt to determine why.) Second, the natural habitat of *Physarum* is the forest floor, where it functions with other organisms in recycling organic material, so its relevance to the ecological monitoring program is clear. Both of those considerations point to this particular study as one of key importance in the program.

Although the research did not involve ecological monitoring itself, effects of ELF EMFs on the variables measured could have important ecological implications. Among the studies in the program, this appears to be the only one in which axenic cultures of a single organism were used. The use of axenic cultures has advantages and limitations. Substantial conclusions about ecological effects cannot easily be made on the basis of a single organism, but effects of an environmental variable are more easily identified and clearly attributable when the study involves a single organism.

SELECTION OF RESPONSE VARIABLES

Laboratory studies of possible ELF-EMF effects on slime mold initiated in 1972 had included three measurements: length of the mitotic cycle, rate of respiration, and ATP content. Mitotic-cycle measurements were included in the original proposal and in experimental results reported up to the 1986 annual report, but were eventually replaced with the ATP determination. Outside data had suggested that ATP content might be a more sensitive indicator of an ELF-EMF effect. Early reports indicate substantial effects on the length of the mitotic cycle, but, according to the 1987 annual report (Goodman and Greenbaum 1988) an increase in the length of the control cell cycle caused researchers to examine their handling procedures. In response to questioning by the committee (E. Goodman, University of Wisconsin-Parkside, memorandum, 1996), the principal investigator stated that "the *onset* of an effect required at least 120 days . . . [and] . . . another 60 days of experimentation was usually performed." Because of "the abbreviated season at the Wisconsin Test Facility, this type of experiment wasn't deemed to be feasible." According to the final report (Goodman and Greenebaum 1990, p. 41), "weather conditions in northern Wisconsin prevented experiments from being carried out for more than 140 days" because cultures would not grow in the cold temperatures from October through May. Only respiration and ATP content were included in the final report. Both respiration (in microliters of oxygen consumed per minute per milligram of protein) and ATP content (in nanomoles per milligram of protein in extracts made in Tris-borate buffer, pH 9.2, 98C°) were deemed suitable for the desired investigation and were examined in the laboratory and at study sites.

Those choices are justified by virtue of the earlier studies, in which effects on the two variables were reported and for which the methods were well established. In addition, they represent basic biochemical and physiologic processes at the cellular level, and any statistically significant effects would have undeniable ecological implications. The same could be said for effects on the mitotic cycle, if they had been included.

EXPERIMENT DESIGN AND IMPLEMENTATION

The use of axenic cultures required an experiment design for the study sites quite different from those of the other projects. Pure cultures were maintained and exposed in polyethylene chambers (7 × 4 × 2.25 in.), themselves placed individually in protective boxes in a hole about 20 in. on each side covered by plywood to protect the system from foraging animals.

Even when placed in the ground near the transmitter, the cells inside the chamber did not receive an ELF-EMF exposure comparable to that which would be experienced by cells free in the soil at the same location. To simulate the ELF EMFs, two stainless-steel electrodes, placed 6 in. apart and about 0.25 in. from the bottom of the polyethylene chamber, were connected to copper collector plates buried in the ground about 1 m from each hole along a line with the predominant electric field. The procedure also exposed the cells to the environmental temperature and to some extent other environmental factors—for example, humidity and barometric pressure—but not others, such as soil moisture.

Two duplicate cultures (chambers) were maintained and exposed at each site; the second served as a backup in the case of contamination of the first.

Biologic Sampling Techniques

Samples were placed at three sites: one control site 7 miles east of the nearest antenna element, one treatment site near the west ground element, and one treatment site near (about 30 ft from) the overhead cables of the antenna (outside the right of way).

Cells were allowed to grow on a solid (nutrient agar) medium in the chambers provided. A portion of each exposed culture was inoculated onto fresh medium each week (subculturing to maintain continuing exposure) at a study-site laboratory. The remaining part of the exposed culture was transported by air to the laboratory at the University of Wisconsin-Parkside, where measurements were made according to the same protocols as used for cultures exposed in the laboratory.

Biologic measurements involved the cellular amounts of ATP and respiration rates. ATP in cell extracts was measured in cell extracts with the firefly luciferase method, which is highly specific and reliable. Respiration was measured with an oxygen electrode, which is also reliable.

On return to the laboratory, macroplasmodia from each chamber were scraped from the agar surface, placed in 125-mL Erlenmeyer flasks containing 25 mL of half-strength growth medium, and incubated overnight with shaking. The final report states that, after washing and resuspension in full-strength medium, ATP and respiration rates were measured, usually within 40-72 h of removal from the exposure sites (Goodman and Greenebaum 1990, p. 5). A similar delay occurred with the cultures exposed in the laboratory, but it was about 9 h less because transportation time was eliminated.

Because the effects of ELF EMFs on this organism previously reported were stated to be reversible, the protocol, which involved maintenance of cells for 2-3 days before analysis, seems to compromise a firm conclusion. If the EMF has only a small effect, ATP content and respiration rates might well return to control levels during the time of study. In response to committee questioning (E. Goodman, University of Wisconsin-Parkside, memorandum, 1996), the principal investigator confirmed that the effect is reversible but stated that "our data suggest that at least 3-4 weeks is required for an effect to dissipate." The final report cites earlier work by the same researchers which found that the lengthened mitotic cycle returned to control levels after 3-4 weeks. However, it seems unlikely that the perturbation of a metabolite like ATP, which turns over rapidly in the cell, would persist long after the conditions are changed.

In any event, it seems that such a problem could have been circumvented by freezing the cells immediately after removing them from the chambers and measuring their ATP content later. When asked about this, the researcher stated (in the same memorandum) that the 2-day period was needed to convert the exposed cultures from macroplasmodia, which were "not conducive to the types of biochemical tests we performed," to microplasmodia. Furthermore, because "the same cultures were used for ATP analysis as well as [respiration] measurements, freezing the cells did not appear to be a viable option." Respiration measurements are made with living cells, and freezing could well have resulted in some damage that would compromise respiration measurements. That would not be so for ATP, however, because it is a chemical substance.

In our opinion, the design of the experiments is problematic. For the procedure to be validated, an experimental control should probably include a repeat of the investigators' earlier reported effects of ELF EMFs on ATP content and respiration rates in *Physarum* showing that the effect did not degrade. To be sure, the protocol for the earlier experiments might also have included the long delay, but it is not mentioned in the final report. The committee wonders whether the variability, which is repeatedly alluded to, might be due in part to the variation in the time from exposure to measurement, assuming that recovery is taking place during that time.

Physical Measurements

Other than ELF-EMF exposure, the only physical characteristic measured was temperature. Temperature was estimated at each site by placing a battery-operated Dickson monitor, calibrated in the laboratory, inside the protective box of one of the cultures at each site. These monitors are accurate to within one degree Fahrenheit and operated satisfactorily except when water got into the chamber. The mean weekly temperature was calculated by averaging the daily high and low at each site; soil temperature was also recorded hourly by IITRI starting in July 1987.

Statistical Methods

Data on both study site cultures and cultures maintained and exposed in the laboratory were evaluated with analysis of variance (ANOVA) techniques. The independent variables used for laboratory data included
 Replication of measurements.
 Time in microplasmodia suspension culture.
 Intensity of EMF exposure.
 Time of EMF exposure.

For study site data, three independent variables were used for the ANOVA:
 Exposure type.
 Exposure site.
 Time of EMF exposure.

Study site data were also analyzed using multivariate linear regression; values derived from hourly measurements of the electric-field component of the ELF EMFs, current density, and temperature were included in this analysis. For each site, these data were averaged over the periods between metabolic measurements. A multivariate linear-regression procedure was used to predict ATP content and respiration rate as functions of the electric-field component of the ELF EMFs, current density, and temperature. The period of EMF exposure was included as a predictor variable. Pearson moment correlations were calculated.

Those methods are probably as good as any alternatives.

Exposure Assessment

Because the currents and ELF EMFs in the individual containers, both in the laboratory and at the study site, were adjusted to mimic conditions at the soil surface, the measured exposures are probably known better for this project than for many of the others. At the study site, the exposures resulted from the operation of the Wisconsin transmitting facility; exposures are well documented in the final report. In laboratory experiments, no attempt was made to duplicate any of the fluctuations in ELF-EMF levels that occurred at the antenna and ground treatment sites.

PRESENTATION OF RESULTS

The hypothesis tested was that exposure to weak ELF EMFs like those produced by the Wisconsin antenna would result in observable physiologic alterations in *Physarum*. The results are presented in the final report in seven tables and 22 figures with commentary that guides the reader to the conclusions and interpretations provided by the authors (Goodman and Greenebaum 1990, pp. 11-40). No statistically significant differences between EMF-exposed cells and control cells in either study site or laboratory experiments were observed.

Tables 1 and 2 of the final report (pp. 12-13) consider the effect of the amount of time that a culture was maintained out of the ELF-EMF environment before analysis, which is related to the issue discussed above—the long time between termination of exposure and measurements. Table 2 indicates that the mean value for respiration decreased with the number of hours elapsed since exposure to ELF EMFs—an apparently statistically significant effect—but no effect on ATP content was found. A concern with the results presented is that Table 2 lists the results of analyses performed after 48, 72, and 96 h in submerged shake culture. If there is a degradation of an ELF-EMF effect, the largest part of it would be expected in the first 24 h. At 48 h, the investigators would be looking only at the tail of that phenomenon. Another worry is that the effect seen, that respiration decreases with time out of the ELF EMF, would mean that the supposed treatment effect is to increase respiration—the opposite of what was reported in earlier laboratory experiments (exposures at higher field strengths).

The authors state that, because of the observed decrease in respiration with time out of the exposure environment, only the 48-h results were used in later statistical analyses. That is suggested as eliminating concern about the issue, which it might not do. In addition, for study-site cultures, it is hard to reconcile that declaration with the description in the methods section, which says that the "analyses were performed within 40 to 72 hours after removing the cultures from the Wisconsin Test Facility." How can the shorter time allow for 48 h in liquid suspension?

For the laboratory cultures, there were unexplained statistically significant differences in the measured values of ATP and respiration rate from year to year. All data before 1985 were omitted from consideration, as is reasonable because the antenna operation was intermittent. But data thereafter also showed differences from year to year for unknown reasons. Study site data on respiration (but not ATP content) are stated to vary significantly with time, and this is discussed by the authors on the basis of aging cultures. There is no comment on how values differ from year to year, but the conclusions section (p. 42) states that "the field and [laboratory] values obtained for *Physarum*'s [rate of respiration] and ATP were similar to those we have previously published."

The authors conclude that there is no indication that the ELF-EMF exposures used resulted in any "extensive bio-effects" (p. 41). They then discuss why this was so if their earlier studies did show an effect. Differences cited include the earlier studies' use of submerged (liquid) cultures for longer times (more than 180 days whereas the current study kept them on agar for about 140 days) and other differences in culture methods. Most important, it seems, is that the original laboratory experiments were performed at intensities 5-10 times higher than those in the current study.

Conclusions and Committee Critique

Validity

These experiments were carried out by an experienced and dedicated team of scientists. The organism selected has a unique place in the evaluation of ELF-EMF effects: it has already been reported as one in which physiologic changes occur after exposure. The research team was expert in the maintenance of cultures and measurement of biologic changes. They worked closely with the team of engineers in setting and recording exposure levels.

Uncertainties

The committee finds no uncertainties in the collection of data, but has a question about exposure and how the data are analyzed in relation to it. The committee also has three major concerns about the design of the experiments. First, as described on p. 25 of the final report, "there are some dramatic differences in the field intensities during weeks 8-9, 11-13 and 17.5-18.5." These were not attributable to the antenna, which was fully operational. On being questioned and in the final report itself, the principal investigator indicated that the reason for the variation was not known but speculated that it might have been due to any of several factors. Such might include differences in exposure cells, differences in the setting of the control potentiometer, changes in growth cell conductivity due to growth of the mold, and local changes in soil conductivity (E. Goodman, memorandum, 1996). The data on times of high intensity do not appear to be treated separately from those on low intensities. Might some different conclusion have been reached if they had been treated separately? Also, exposure data are plotted starting in the middle of week 6, whereas ATP content levels are shown starting at week 2. What exposure values were used for those early weeks?

Second, if ELF EMFs have an effect on the respiration rate of cells and on cellular concentrations of ATP and the effect is reversible (as it is stated to be on the basis of earlier results when effects were recorded), the return to the original state should begin at the time that exposure to ELF EMFs ends. In going from state A to state B in a model system, the rate of change usually is maximal at the outset and the final state is approached asymptotically. Supposed effects in this system are not measured until much later—between 40 and 72 h after the exposure is stopped, according to the methods section, or after 48, 72, and 96 h in liquid suspension culture, according to the results section. That compromises the conclusions substantially. The authors do not discuss or refer to the fact that ATP is a product of respiration and that the two are thus expected to be related.

Third, ELF-EMF intensities only 0.1-0.2 times those previously shown to have an effect were used. As already noted, the use of this organism was of special value because of the earlier studies, but those experiments were not duplicated. However, the authors still refer to them as valid and correct. It can be argued that the objective of the monitoring program was to determine whether the emission of the Wisconsin facility would cause any physiologic effects and that higher intensities were therefore not relevant. Such a view is invalid; if there is an effect at a higher intensity, the most-important objective should be to determine the nature of that effect, from which one can evaluate the possible effects at the lower intensity. Moreover, it is widely agreed that there are variations between individual organisms and between species. Some test subjects are killed by doses considerably lower than the dose at which 50% of the subjects die (LD_{50}), and there are species differences. Slime mold might not respond to fields only 0.1 times those of the original study, but some other species might be responsive at such fields. Any effect at any dose is of potential interest because it could lead to an understanding of mechanism.

Another consideration is that in experimental science one of the major techniques for establishing cause and effect is to determine dose-response relationships, but this was not done here. The researchers have measured only one point on a putative curve—an unsatisfactory experimental approach—and the response to ELF-EMF exposure reported earlier was not replicated. In the ecological monitoring program, the technical aspects of imposing fields were presumably well managed, so it would be good to verify the earlier results and proceed from them. That has not been done, and without a dose-response relationship it is difficult to say that there is no effect.

WISCONSIN BIRDS AND MICHIGAN BIRDS

Project Proposal

The Navy's original monitoring plan called for research on bird populations that focused on potential effects of ELF EMFs on bird migration and orientation (IITRI 1976, pp. 37-41). Those subjects were given priority because existing data indicated that at least some birds use geomagnetic fields for orientation and navigation. Several kinds of monitoring programs were suggested. The two given the most attention were radar tracking to assess the number and trajectory of birds migrating near the antennas, from which inferences could be drawn about whether birds are disoriented by antenna-generated EMFs or actively avoid them, and conventional ground counting to determine whether breeding migratory birds avoid ELF EMFs or recruitment of new breeding birds is reduced over the years because of effects on reproductive success. An additional suggestion was longitudinal studies of marked individuals to make inferences about effects of the antenna on survivorship, reproduction and site fidelity; this approach was never seriously pursued, because it requires impractical sample sizes.

Initially, IITRI accepted a proposal (dated 1982-1983) that emphasized radar tracking to determine effects on migrating birds. The proposal also included some ground counting. Peer-review comments in 1984 criticized the design and execution of the project, and it was terminated.

A proposal from G. J. Niemi and J. M Hanowski, of the University of Minnesota, Duluth, dated May 1984, was accepted as a replacement. It emphasized ground counting of birds. The investigators proposed to use ground censuses along sample transects to determine whether there were differences in species richness, relative density, and relative frequency of breeding birds between treatment areas adjacent to the antenna and control areas away from the antenna. The proposal recognized that four factors needed to be accounted for in an analysis of variance in various abundance measures: habitat type, region (Michigan or Wisconsin), ELF-EMF intensity, and right-of-way (ROW) clearing effects.

Region was handled by separate analyses in Wisconsin and Michigan, as is appropriate. Habitat type was handled in Wisconsin by statistical models. In Michigan, attempts to match habitats on treatment and control sites resulted in all the control sites' being clustered to the southeast of the antenna. That creates what is known as lack of interspersion and results in a statistical flaw known as pseudoreplication (see Chapter 4 for a more-detailed discussion). ELF-EMF intensity was used to establish treatment and control sites; but in later analyses of the response variables, year served as a surrogate of ELF-EMF intensity. ELF-EMF intensity varied within and across years, so the power of statistical tests is difficult to assess. In dealing with ROW clearing effects, the design gambit was to eliminate "[the ROW] factor from the sampling design by eliminating this zone of habitat [near the ROW] from sampling." (Niemi and Hanowski 1984, p. 2). Treatment sites were 125 m from the edge of the ROW clearing and hence 150 m from the antenna. Consequently, the high-intensity treatment sites had magnetic field intensities less than 0.1 times those under the antenna.

The switch in project teams shifted the focus of bird population studies from direct, immediate impacts on bird navigation and migration to indirect, potentially delayed impacts on local bird abundance.

Species Selection

The investigators were not intentionally selective about bird species. They counted all birds detectable from line transects, using conventional transect methods. The approach is well-established, standard procedure. The technique is most sensitive to conspicuous birds, like singing adult males, and least sensitive to inconspicuous birds, like silent, subordinate young or females on the nest. Although different species, sexes, and age classes are differentially detectable and habitat differences influence detectability, it is unlikely that these considerations would contribute to artifactual treatment effects in the statistical analysis of this particular data set.

Response Variables

The response variables chosen were abundance indexes of individual species and aggregate indexes of species richness. Local abundance results from the integration of demographic processes (birth, immigration, death, and emigration), some of which can operate on fairly long time scales in long-lived organisms. These abundance indexes are basically sound for assessing effects that involve birds' perception and avoidance of EMFs. The indexes would have little power to detect small demographic changes with effects on abundance that involve long lag times. For instance, if adult breeding birds were recruited primarily from locally produced offspring and nestling production decreased by around 10%, the effects on bird abundance would be severe in the long run but might not be detectable by the end of the study, because of the slow turnover of adults. Other projects on nesting success of tree swallows (see "Small Vertebrates" below) are aimed at these kinds of effects. Another possibility is that demographic rates compensate to maintain local abundance. For instance, birds that die or depart after long exposure might be replaced by cryptic subordinates ("floaters") from the surrounding area. If the EMF changed the treatment sites into demographic "sinks" without dramatically lowering population densities, that would be missed.

Experiment Design and Implementation

The method of walking along transects and counting birds by sight and sound is well established. The investigators used great care and sophistication. Daily and seasonal timing were appropriate to the objectives. The location of the treatment transects, 150 m from the antenna, is problematic.

It is well known that ROW clearing through forest habitat affects bird populations, but the exact mechanisms and magnitudes over distance from the edge are ill understood. EMF strength and ROW effects on habitat are greatly confounded. The treatment transects were placed parallel to the antenna at a distance of 150 m ("near" the antenna) in the hope that this would "allow us to separate effects of electromagnetic fields on bird species and communities from effects due to direct habitat changes along the ROW" (Hanowski et al. 1991, p. 1). The investigators chose to have the treatment transects at 150 m with the knowledge that magnetic fields at that distance were still more than 10 times as high as magnetic fields at control transects and in the hope that any unmeasured ROW edge effects were negligible. The 10:1 criterion was established by IITRI.

An unfortunate consequence of that gambit is that the magnetic field in the air at the treatment transects, although more than 10 times that of the control transect, is less than 0.1 times the magnetic field under the antenna (see Table G-6 of the 1989 engineering report, where magnetic flux at 150 m is less than 5% of the flux under the wire). Figure 3-1 shows the distance decay function for magnetic flux. The treatment transects are off the hump and out on the tail of the magnetic field. It is apparent that, given the IITRI 10:1 criterion, the treatment transects at 150 m could just as well be called control transects for a hypothetical treatment transect under the antenna.

Although they could often detect birds up to 100 m from the transect, the investigators' sensitivity must have been relatively low at ranges over 50 m. Consequently, comparisons between treatment and reference transects are testing for effects of the combined ELF EMF and ROW clearing at roughly 50-250 m (more probably 100-200 m) from the wire. Either effect, if it exists, should be greatly diminished at this distance.

In principle, ROW edge effects and ELF-EMF effects could be decoupled temporally in Michigan, where there are data on conditions after ROW clearing but before antenna activation. Treatment transects could have been placed nearer the antenna to determine the incremental effects of the field, beyond ROW effects. Even so, if ROW edge effects developed slowly, they would still be confounded with the onset of antenna operation. Proper design would also require the establishment of matching ROW effects at the control sites.

There is no easy solution to the problem. At this point, it is essential to make explicit how the experimental design limits the interpretation of the results. The generally negative results do not show that EMFs associated with the Navy's ELF antenna have no effect on bird populations. They indicate no consistent, measurable effects beyond roughly 50-100 m, where magnetic fields in the air have dropped to less than 10% of their magnitude under the antenna.

Although control sites were selected by a seemingly adequate randomization procedure, all the control sites in Michigan were southeast of the antenna and treatment sites (Hanowski et al. 1994, Figure 1). That might have resulted from efforts to relocate control sites so that they would better match the habitats found on the treatment sites. In any case, treatment and control sites were not interspersed. Less important, treatment transects were oriented mostly north-south and control transects mostly east-west. In Wisconsin, treatment and control sites were not interspersed, but the controls at least surrounded the antenna, and compass orientations were more varied within groups. The inability to intersperse treatment and control sites "at random" raises subtle questions about pseudoreplication, but it does not seem to lead to obvious misinterpretations of the results.

Physical Measurements

Seemingly adequate measurements of habitat structure were used to establish statistical control (Wisconsin) or to match habitats between treatment and control sites (Michigan). Physical factors, such as microclimate, were assumed to be randomized across sites.

Statistical Methods

The investigators used repeated-measures analysis of variance (ANOVA) and analysis of covariance (ANCOVA) (to incorporate habitat differences in Wisconsin) on species abundances and aggregate indexes over time on treatment and control transects. The between-subject factor was location (treatment versus control), and the within-subject factor was year. It was expected that abundance indexes would fluctuate from year to year. The main question was whether these fluctuations were different on treatment and control sites. Statistically significant interactions between factors would indicate that changes in bird abundance over time were not similar in treatment and control areas. The plan was to use the technique of multiple contrasts to incorporate informa-

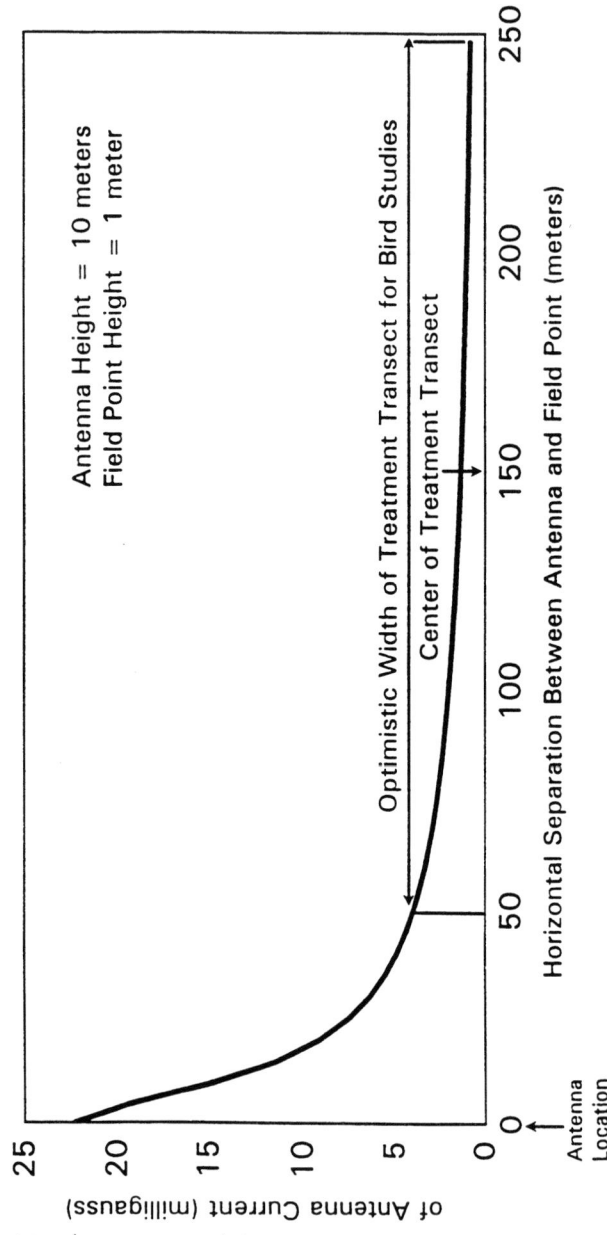

FIGURE 3-1 Magnetic field is inversely proportional to separation from the antenna.

tion on antenna power levels if the screening based on years turned up statistically significant interactions. The strategy of screening for treatment-by-year interactions in an ANOVA (with repeated measures over years) seems reasonable, but the somewhat erratic pattern of antenna power within and across years raises concerns. Information on ELF-EMF strength was not used effectively. Changes over time unrelated to antenna operation, such as localized drought or localized insect populations, could lead to false-positive results (type I errors) if they occurred at either the treatment or control site, but not both. The possibility is increased by the lack of interspersion between treatment and control sites, and it seems to have happened in Michigan (see, for instance, Figure 6 in Hanowski et al. 1994). Several statistically significant treatment-by-year interactions are dismissed because they appear to be unrelated to antenna activity. False-negative results (type II errors) could result if variations in power make year a weak surrogate for ELF-EMF strength. Inspection of the data suggests that this is not a problem with this data set. Some sort of formal before-and-after, control-and-intervention (BACI) analysis would be more appropriate.

Although the investigators conscientiously estimated the power of their statistical tests, it is impossible to estimate their power accurately because of the somewhat messy relationship between year and field strength and the pseudoreplication problem.

Quality Assurance, Quality Control, and Replicability

The quality-control and quality-assurance efforts seem to exceed the general standards for this sort of work. Peer reviews were generally very positive about the yearly reports, and the investigators addressed the few questions that were raised. Questions about habitat differences across treatments in Michigan resulted in relocation of control sites to match treatment sites better (investigators attempted to deal with this problem in Wisconsin statistically, by using ANCOVA to incorporate habitat differences explicitly). The severe lack of interspersion might have been due in part to this relocation; investigators might have had to cluster their control sites in this relatively restricted area to find habitats that matched the treatment sites.

It is impossible to gather this sort of observational data "blind." One has to trust that investigators do not consciously or unconsciously introduce biases. These investigators have made a conscientious effort to reduce the risks by using experienced field workers and rotating them between treatment and control locations.

Exposure Assessment

IITRI made adequate spatial maps of ELF EMFs at full-power antenna operation. It is impossible to determine the exposure levels for highly mobile adult birds. The relationship between the magnetic-field strength and distance from the antenna (as shown in Figure 3-1) means that the average exposure is higher than that expected at the average distance, but average distances for actively foraging adults are unknown. The hypothesis tested was not about exposure effects themselves, but about changes in bird abundance in the space near the antenna, where ELF EMFs are high, at least on the average, compared with that in space farther away. Although it is not essential that individual dosages be known, better spatial maps of average fields integrated over time would have been useful.

RESPONSE TO REVIEW

IITRI responded to the critical reviews of the initial project by canceling it. Peer reviews of the replacement project were generally very positive. Reviews of the first annual reports from this project expressed concern that ROW effects on habitat were not eliminated. None of the peer reviews expressed concern about the low intensities of the magnetic field at the treatment sites.

PRESENTATION OF RESULTS

The reports are very well written, facilitating critical review. With the few exceptions already noted, each decision is clearly explained and justified. The analyses and conclusions are explained clearly. Reasonable alternative hypotheses are raised and adequately considered.

CONCLUSIONS

Many bird species and several aggregate community indexes were compared across locations and years and subjected to statistical tests. About as many comparisons were statistically significant at the conventional level of 0.05 as one would expect to occur by chance. In addition, apparent false-positive results were considered case by case and dismissed. The investigators interpreted the results reasonably: "We found no convincing evidence that overall breeding bird distribution or abundance was affected by electromagnetic fields produced by the ELF antenna" (Hanowski et al. 1994, p. 27). That is a reasonable interpretation, as long as it is kept in mind how far from the antenna the investigators were looking for effects and the potential insensitivity to demographic impacts that might take a long time to become detectable in abundance effects. The conclusion should not be generalized beyond the low magnetic-field strengths found at the distances at which censuses were taken (less than 10 mG and less than 10% of the value beneath the antenna).

An analogy might help to reinforce this message. Imagine that the magnetic-field strength function shown in Figure 3-1 was actually a cross section of a hillside and that one is interested in knowing whether the hill had an effect on the density of birds in the landscape. Then, imagine that we found that the density of birds living 100-200 units from the peak was about the same as the density more than 1,000 units away. One could not conclude that the hill has no effect on the population, but, one could conclude that no effect on the index of abundance is apparent across low elevations. One could also reasonably conclude that there are not widespread effects over the landscape, at least in the short term. Any effects that do exist, even if locally severe, are concentrated in a relatively small area.

There are always questions about how far results like these can be generalized. The goal of the project was not to discover generalities about birds and EMFs, but rather to assess the impacts of particular transmitting facilities on bird abundances. It is always possible that even at the ELF sites seemingly small effects on populations, mediated through modest reductions in reproduction and recruitment, will slowly compound and only later become apparent. At present, there is no evidence of statistically significant, widespread, short-term impacts of EMFs associated with the ELF antennas on bird populations. The studies would not have detected any effects within roughly 50-100 m of the antennas.

SMALL VERTEBRATES

Project Proposal

This project attempted to ascertain whether the operation of the Navy's ELF antenna system in Republic, Michigan, had effects on small vertebrates—chickadees, tree swallows, chipmunks, and deer mice—in an ecological context. It is clear that even a partial answer to this question could have exhausted all the resources allocated to the entire program. The investigators used their scientific judgment and the state of the art in research on EMF biologic effects to select four species and a few important variables to examine, and they divided their overall effort into six substudies that were completed:

- Fecundity, mortality, growth, and maturation in tree swallows.
- Growth and maturation in deer mice.
- Homing in tree swallows.
- Homing in small mammals.
- Development in tree swallows.
- Maximal aerobic metabolism in chickadees and deer mice.

Three other planned substudies were eliminated in March 1989 because of budgetary constraints associated with a negotiated increase in the salaries of nonfaculty employees: on small-mammal communities, small-mammal parental care, and tree swallow incubation.

FECUNDITY, MORTALITY, GROWTH, AND MATURATION IN TREE SWALLOWS

This large substudy examined several aspects of reproduction in tree swallows, including number of eggs per clutch, hatching success within a clutch, fledgling success, rate of growth and probability of mortality of eggs, young, and nests.

It was a generally well-conducted substudy with meticulous attention to experimental detail. Some of the variables were direct measures (such as, egg weight and volume), and others were indirect measures that resulted from fitting the data to a model function and then comparing fitted parameters in the model function across conditions. The substudy also included a creative nestling-swapping experiment (in 1990 and 1991 only) to separate the effects of EMF exposure during the period in ovo and compare it with the effects of exposure of nestlings. A number of control and sham permutations were included in the designs.

The investigators concluded that no effects were consistently attributable to the operation of the ELF antenna system.

Three major questions arose in this substudy. Two—statistical power and integration of exposure data—are common to the entire set of substudies and are discussed below. The third, peculiar to this substudy, was related to the nonlinear least-squares fitting technique used to fit the growth data to nonlinear functions. The investigators used the NONLIN module of SYSTAT, which contains robust implementations of a steepest-descent (also known as the Newton-Gauss method) minimization method and the slow but relentless simplex method. The investigators do not provide details, so it is not clear which method they used. It is also not clear whether they tested various starting conditions in parameter space to ascertain whether they had reached the true minimum or the algorithm had been "fooled" by a local minimum. Trapping by local minimums is a problem that can occur with either algorithm, but the steepest-descent methods are particularly susceptible to this problem. One wonders, when one sees isolated large differences in mean values (as, for example, in the 1985 test in Figure 17 or the 1988 test in Figure 64), whether they reflect true variability or an artifact of the fitting process.

GROWTH AND MATURATION IN DEER MICE

This substudy terminated in 1991 because of technical problems. The design was similar to that of the substudy on avian growth and maturation. It produced one of the few positive effects that is not "explained away" in the general discussion: a pattern of earlier eye-opening in deer mice from test plots during periods of full operation than from control plots. The meaning of that effect, if any, for later development, health, or reproductive fitness of deer mice is unclear. All other measures were interpreted as negative by the investigators—a reasonable conclusion in light of the few data that were obtained.

HOMING IN TREE SWALLOWS

This substudy addressed the homing success of tree swallows by measuring the proportion of animals that, on displacement, successfully returned home and the time required for the return, which was around 4-5 hours. The flight paths for test-plot and control-plot animals were of comparable length and required the animals to fly over one leg of the antenna system.

This was a carefully conducted substudy. The investigation paid attention to findings that could result from possible artifacts due to peculiarities of the release site. The investigators observed differences in success and speed of homing that depended on the origin of the birds (from control or test plots).

They carefully examined whether the differences, which were seen every year, were due to plot differences or antenna operation, and they concluded that antenna operation was not related to the observed differences.

This substudy exemplifies the internal inconsistencies that arise when the distinction between short-term and long-term effects is ignored and the exposure-assessment data are not integrated with biologic design considerations. Although the possibility of short-term and long-term effects is acknowledged (final report, p. 150), the acknowledgment does not translate into a clear separation into two hypotheses that require separate experiment designs. That is evident in the tortuous reasoning (pp. 152-155) that leads to the investigators' conclusions that the antenna system did not have an effect on homing behavior; their reasoning freely mixes potential short-term and long-term effects. The investigators also did not indicate which of the following three distinct hypotheses they were testing, their statements on p. 150 notwithstanding:

a. If the hypothesis is that the long-term exposure to the electric and magnetic fields produced by the Navy ELF system causes a long-term effect in homing ability in birds, it is not relevant to have the birds fly over the antenna system; there should be differences between test-plot birds and control-plot birds released from several sites where the flight path does not intersect the antennas.

b. In contrast, if the hypothesis is that test-plot and control-plot birds are comparable (that is, there are no long-term effects) and the homing effect results from antenna operation, it is less important whether there are plot differences, and it is all-important to ascertain whether the antenna leg that intersects the flight path was active during the 4-5 hours that the birds were in flight. Even in years of full antenna operation, there were down periods, and in some cases the legs of the antennas were rotated in their activity.

c. Finally, if the hypothesis is that long-term operation of the antenna system causes some long-term change in homing activity in birds and that the effect is altered or magnified by acute exposure resulting from flight over the antenna, the experiment design that was used is appropriate. However, it is still important to ascertain whether the antenna leg that intersects the flight path was active during the 4-5 hours that the birds were in flight. Not only could that lead to exposure misclassification (and result in a bias of the results toward the null hypothesis), but it would confound attempts to clarify whether the observed, consistent differences between test-plot birds and control-plot birds are the result only of plot differences or of interaction between plot (chronic exposure) and antenna activity (acute exposure). In Table 39, the investigators identified (footnotes C and D) some conditions in which they knew that the antenna was on for only part of the flights and that the proportion of time that it was on varied with each individual bird. In addition, they acknowledge (p. 150) that the antenna operation is highly variable, nearly hourly.

HOMING IN SMALL MAMMALS

This substudy was patterned after the tree swallow homing study but used deer mice and chipmunks. Considerable creativity was exercised in what must have been a difficult project. Although the data are generally consistent with the conclusion that the antenna system did not affect homing behavior in deer mice and chipmunks, the investigators acknowledge (pp. 163-164) that the sample sizes were often inadequate. For that reason, this substudy cannot be considered informative.

DEVELOPMENT IN TREE SWALLOWS

This substudy characterized the normal embryologic development in tree swallows. Except for the common questions of statistical power and of whether the antenna was active in the period when the eggs were in the nest before collection it is an excellent substudy. The investigators used the well-accepted developmental framework provided for the chick by Hamilton and Hamburger in their landmark compendium and staging, and they adapted it to the different developmental timeframe of the tree swallow. They used accepted, noncontroversial histologic methods and provided good rationales for the judgments that visual evaluation of embryos always entails.

The results are generally negative, although a few isolated points reached statistical significance. However, the investigators' argument that the pattern of changes is neither consistent nor compelling for an antenna effect is convincing. Some specific aspects of their discussions are not so straightforward. In their discussion and comparison with other species, they suggest that differences in waveforms between air and the egg-mother environment might have played a role in the general confusion present in this literature. They refer to work by Martin, who deliberately exposed eggs to different waveforms. Although the investigators might have a case with respect to electric fields, the waveform from a magnetic field will not be altered by the egg-mother environment. Far more relevant to a comparison with other species is other work by A. H. Martin (Martin 1989; not cited in their report) that showed, in the same laboratory under the same conditions, consistent differences that were directly attributable to which strain of chick eggs (either Arbor Acre or White Leghorn) was used in the experiments. If the use of different strains of chicks can lead to different results, why not the use of different species?

MAXIMAL AEROBIC METABOLISM IN CHICKADEES AND DEER MICE

This substudy sought to measure the peak metabolic rates (PMRs) of chickadees and deer mice during winter. The authors used the "helox" technique, which takes advantage of the higher thermal conductivity of helium than of nitrogen. For each measurement, the animals are given a helium-oxygen mixture to breathe while they are held at a preset low temperature. The helox mixture causes an increased loss of heat from the body, which, in turn, triggers an acute thermoregulatory response. The motivation of this approach appears sound ecologically, although the enthusiasm for this single measure as an index of "physiologic health" and ability to cope with "stress" appears to be more widely accepted by the authors of the quoted textbooks than by the entire community of physiologists.

The substudy is problematic. Thermoregulatory responses cannot be equated with "stress" in the broad sense, in that, depending on the species, thermoregulatory responses might not share the hormonal profile of stressors. For example, the general stress response in most mammals involves the ACTH-adrenal steroid axis and sympathetic activation leading to releases of norepinephrine. A thyroid hormone response is sometimes but not always found. In contrast, the thermoregulatory response in birds is associated with an absence of the norepinephrine response and the main hormonal mediator is thyroid hormone.

Even if one considers PMR on its own merits, the measurement must be done carefully. As outlined by the investigators (p. 189), a true peak can be ascertained only by taking several measurements and showing a dropoff at temperatures above and below the peak. The peak temperature must be determined for each individual. The investigators tested each temperature only once a day and needed a minimum of 3-5 days to determine the PMR. Adaptation to acute temperature stresses has been reported, although it is not clear whether that presents a problem in this context.

A more important problem is that the investigators were not able to obtain complete, high-quality data on all the animals. They had to develop a rating scheme for "data quality." In addition, although 86-88% of the deer mice were able to complete all measures at acceptable quality, the success rate in chickadees was only 37-40%. Of the birds, 16-20% died during the initial transition to captivity, and an additional 34% of the ones that were left died during the multiday testing procedure. A physiologist wonders about a procedure to examine "physiologic health" in ostensibly healthy animals that results in the death of one-third of the animals.

A number of analyses revealed plot differences and operation-period differences but not plot-operation interactions. On the basis of the analyses and the feeling that only the high-quality data classes should be considered definitive (p. 205), the investigators concluded that no differences could be attributed to antenna activity.

COMMON LIMITATIONS OF SUBSTUDIES

Integration of Exposure Assessment Into Experimental Biologic Design and Statistical Treatment of Data

An important limitation in all the substudies was the failure to integrate the exposure-assessment data into the biologic design. It is regrettable because most of the substudies collected data carefully. The limitation seems to have arisen in part because of a lack of broadly EMF-knowledgeable biologists at the project-definition and project-overview stages. This is also true for the other projects in the program. Although many publications have arisen out of the Navy's ecological monitoring program, the only two investigators that had and continue to have distinguished positions in the EMF-effects community are E. Goodman and B. Greenebaum.

At the time of the 1977 National Research Council report (NRC 1977), it was known that the exposure metric that might be relevant to biologic responses to EMF, if any such responses exist, had not been ascertained. No one had been able to define the biologically relevant "dose" metric. That situation continues. Some investigators examine the time-weighted average field exposure (TWA), others peak exposure, and others intermittency in field exposure as candidates for the biologically active field metric.

Also unknown is the duration of exposure that is required for a response of an organism or the time required after exposure ends for the effects of EMF exposure, if any, to disappear; the relevant time constants have not been defined. However, that does not mean that it is impossible to conduct useful experiments or that specific hypotheses about exposures and responses cannot be articulated and tested. The series of experiments presented in this final report (except the homing experiments) failed to ask that question, so the experiment design and statistical analyses are muddled and, in some cases (as in the PMR work), internally inconsistent.

In particular, the statistical analyses use, as a measure of "dose," a three-level stratification of exposure based on the fraction of time that the antenna system was operational. That has several implicit assumptions that are seldom made explicit. The arbitrary definition of dose, which assumes that the fraction of time that the animals are exposed to any ELF EMFs is the relevant variable, implies little dependence on the level of operation (which for magnetic fields is proportional to the current in the antenna). If the investigators believed that level of antenna activity was important, the obvious metric would be the product of the time of exposure and the field intensity; this was not used.

It should be a truism that the times of operation of the antenna are important only if they coincided with times when the biologically relevant processes were taking place. For example, in the developmental studies of tree swallow eggs, it does not matter whether the antenna was on or off for the entire year, but only whether the antenna was active while the eggs were developing inside the mother or during the 3-4 days after the appearance of the last egg, when the embryos were being incubated before the clutch was collected and analyzed. Consider the following three scenarios:

    a.   For clutch 1 in 1988 the antenna was active for 2 of the 4 days of maternal incubation during this level 1 year.
    b.   For clutch 2 in 1988, the antenna was active for 4 of the 4 days of maternal incubation during this level 1 year.
    c.   For clutch 3 in 1991 the antenna was (because of repairs or other downtime) active for 0 of the 4 days of maternal incubation during this full-operation year.

In those scenarios, the exposure assignment should be

    Clutch 1: level I (as defined by the researchers).
    Clutch 2: full.
    Clutch 3: inactive.

Instead, the classifications based on the investigators' exposure system are

Clutch 1: level I.
Clutch 2: inactive.
Clutch 3: full.

The resulting exposure misclassification tends to bias the results, in general, toward the null hypothesis (no effect), whether that is true or not. This is potentially correctable in that the investigators know when they collected a specific clutch and the Navy knows when the antenna was active. There is no evidence that this question was asked.

The analysis by Beaver et al. (1994) intrinsically also assumes that the biologic time constant for onset of an effect is instantaneous and the constant for decay of an effect essentially infinite. That leads to the following internal inconsistencies in the experiments:

    a. By the design of the PMR studies, the animals (from control plots and test plots) were already acclimated to the winter temperature. The helox procedure used to measure maximal metabolic rate can define the acute response in already-acclimated animals. Presumably, the hypothesis is that EMFs from the ELF antenna system could impair the acute response to cold, which is mediated primarily by a release of norepinephrine in mammals but not in birds, in which the main mediator is output of thyroid hormone (for example, see Prosser 1973). However, it must also be assumed (it is not stated) that EMF exposure affects *chronically* only the *acute* response and not the acclimation process. If it did not affect the acclimation process, the animals from test plots and control plots would differ in metabolic capacity on the day they were caught, but no data were collected on that day. Why assume that EMF exposure chronically affects a fast, transient, minute-to-minute thermoregulatory mechanism (acute response to cold stress), and not also test whether the EMF response in this physiologically fast process might itself be fast and transient? Why not assume that (or at least test whether) nearly-continuous EMF exposure (during the full-operation years) affects the acclimation process?

    b. Along the same lines, the animals were held in a holding area for at least a day before any measurements were conducted. The investigators report performing only one measurement per day and holding the animals for 2.5 weeks for retesting. Under the infinite-decay-constant assumption, the procedures are appropriate and the results obtained are consistent with the hypothesis: there are no differences in retesting measurements from short term to long term. But if a metabolic effect had a 1-day decay time constant, which is not an unreasonable value for active physiologic responses known to result from magnetic field exposure (for example, measured activation and decay of ODC in vitro or alteration of heart-rate variability in humans), then the animals from both control and test plots would have decayed back to baseline by the time of the first measurement, and no differences would be found. The comparisons of first-day rates to attempt to analyze short-term effects would already be too late to detect short time constants, and the considerable amount of work in the comparisons would be uninformative.

Experiments with Statistical Power Too Low to Yield Meaningful Information

The investigators decreased the sample size, and hence the power of the studies early in the process. In fact, the power that they settled for (a 70% chance of detecting a difference of a given magnitude) is below biologic standards. That makes it much less likely that effects, if any were present, would be detected. As a project peer reviewer—who was otherwise favorably impressed as he followed the project from year to year—notes, it is not clear that the project would have been found meritorious enough to be funded if the lowered power had been proposed from the start. When combined with assumptions about information that would have been obtained in the projects that were dropped, it is clear that the information obtained is severely limited and statistically biased toward not finding an effect even if one was present. That is important for the evaluation of this program for the stated goals of the Navy's ELF monitoring program, and it raises questions about the management of the project.

To allow investigators to drop the nominal power of some of the data collected, and hence of the statistical tests, to 70% is distressing; to allow designs that lead to statistical power of 30% or even less is unconscionable (see, for example, Beaver et al. 1994; pp. 49-51, swallow fecundity; pp. 72-73, swallow body-mass growth and age at maximum; pp. 82-84, swallow tarsus-length variables; pp. 92-93, swallow ulna; pp. 93, 96, swallow wing; pp. 102-103, swallow landmark events; pp. 120-123, swallow growth in relation to electric-field or magnetic-field intensity; and pp. 137-140, deer mouse landmarks). An accurate and sober appraisal of the consequences of this approach were clearly spelled out by a peer reviewer for the small-vertebrates project (letter dated February 15, 1990, p. 2, first paragraph):

> Approach Used in Statistical or Modeling Efforts. The statistical treatment of the data continues to be excellent, although the need to drop the

power of the treatment from 90% to 70% for some data is sobering. While the reason for doing so is clear, it raises a question as to whether or not the final data will be effective in resolving potential ELF effects? *I doubt that the studies affected would have been endorsed had their original standard for statistical sufficiency been at the 70% certainty level.* We'll just have to wait and see." [emphasis added]

If the peer reviewer was disturbed by a drop from the 90% power "gold standard" to a goal of 70%, what are we to make of designs and data with statistical power of, in some cases, less than 30%? In epidemiology and in clinical trials, where subject participation can be problematic (an approximate human equivalent of animal field trials), the standard is often relaxed to 80%. Below that, serious questions are asked as to whether a study is worth doing, as the peer reviewer questioned.

The reason why studies that have low power and show no effects are not considered informative is that their results will fail to exclude the null hypothesis in any useful way. For example, if a specific study with a 50%-power design indicates that an effect of a 35% change in a variable was not found, there is an essentially even chance that a repetition of such a study could find an effect of up to 35% and a hard-to-quantify but disturbingly nonzero chance that another repetition of the study could find an effect of more than 35%. There are rigorous ways to quantify such probabilities. These methods have become available to the nonstatistician only recently with the advent of inexpensive computing power, and a research team should not be faulted for incorporating in the early 1980s a design that calls for early 1990s computing power. Where their judgment is called into question is in encouraging and performing studies that have been known since at least the early 1960s not to be informative.

An alternative, nonmathematical way to illustrate the ramifications of studies with low power is to consider that the easiest and cheapest way to conclude that an effect is not present, even if it exists, is not to look for the effect ($0 spent). The next easiest way ($X spent) is to look for an effect in a cursory fashion, so that only a huge effect is likely to be observed. The situation does not change substantially if, instead of one variable in a cursory fashion, five variables are examined in an equally cursory fashion ($5X spent). The only conclusion possible is that for each of the five variables no huge effect is likely to be present, but little or no confidence attaches to the failure to observe an effect.

Finally, if we say that $5X is all the money that is available, we can either conduct five cursory examinations or put all $5X into the examination of one variable with appropriate power. The last approach has as a resource-limited conclusion that there could be an effect on any of the potentially infinite variables that were not examined. It also has the conclusion that the one variable that was examined properly could be categorized as showing "effect" or "no effect" with a high degree of statistical certainty. That is the approach that is almost universally applied in scientific research—but surprisingly not by the IITRI team. Scientific data are generally used only to accept a null hypothesis of no effect or to rule out the null hypothesis because an effect has been demonstrated. Hence, it makes sense to make the "effect-no effect" decision with the highest possible degree of statistical certainty.

To the extent that some of the substudies produced data with statistical power ranging from 70% to less than 30%, such data cannot be considered to contribute to ascertaining whether the Navy's ELF antenna does or does not have an effect on neighboring small vertebrates.

CONCLUSIONS

The investigators distilled a number of questions of interest into hypotheses that were testable, at least in principle. Except on the studies of maximal aerobic metabolism, it is difficult to argue with their rationale and choice of variables to measure. Their choice of species seems reasonable and is well justified. They implemented generally sound programs to test their hypotheses. Apart from the lack of an appropriate and hypothesis-driven integration of the ELF-EMF exposure data with the biologic questions, their designs were generally good and the statistical analyses appropriate. However, several of the statistical analyses would have been different had the experiment design taken into account the complexities of EMF dose assessment. In general, the data from a given year were pooled and compared with those from other years; there was little attempt to ascertain the replicability of a given finding within the study except on a year-by-year basis.

The investigators were generally responsive to the peer reviewers, one of whom submitted long, detailed critiques and appears responsible for several improvements and for the increase in the quality of exposition over the years. This was a multipart, complicated project, and the annual progress reports reveal the "growing pains" of the investigators, who sought and eventually found a format that was informative and readable. The presentation of the results is generally clear, and the discussion of the confounding variables that were identified was satisfactory.

The peer reviewers consistently commended the investigators for their attention to detail and for their candor in recognizing problems and pitfalls and their coming up with reasonable and constructive alternatives. However, after ascertaining that the variances of several of the variables of interest were higher than had been estimated, the investigators proposed (and apparently received IITRI approval for) reducing the statistical power of their tests from 90% to 70%, instead of discarding some projects and increasing the sample size of the remaining projects to be able to meet the normal biologic standard of 80-90% power. In fact, when all the data were analyzed, many tests had powers of 30% or less.

The investigators, after carefully analyzing their data and controlling for seasonal, yearly, and plot differences, reached the overall conclusion that the operation of the ELF antenna system did not lead to a consistent pattern of changes in the measured variables. There appears to have been some effort at integrating various findings and interpreting occasional positive, statistically significant findings as negative (for example, see the discussion on pp. 125-126 of the final report). There was little consideration of alternative scenarios, for example, that because of lack of statistical power the null hypothesis could not be accepted or rejected. Another alternative scenario that was not considered was that biologic effects of EMFs might not have monotonic dose-response relationships. If the latter hypothesis had been considered, the investigators might have been struck by how often the values obtained at the intermediate level of antenna activity differed from those at both the inactive and the fully active states, which were similar to each other (V-shaped dose-response curves).

Even if the protocol had included a sound integration of exposure and biologic data, the limitations with respect to statistical power would lead to serious questions as to the appropriate extent of confidence in the investigators' conclusions. Because of these limitations, it is not clear how informative the study results were in answering the basic question of the ELF monitoring program. Some of the additional analyses proposed in Table 5-1 might yet provide informative results.

LITTER DECOMPOSITION AND MICROFLORA

Project Proposal

Litter decomposition and associated microbial population dynamics control the flow of carbon and nutrients through soil, the ability of plants to take up nutrients through mycorrhizal symbioses, and the transmission of root pathogens. Accordingly, this phase of the ELF monitoring program attempted to detect whether ELF EMFs from the Navy ELF communications system in Michigan altered those processes. If it did, there could be long-term consequences for plant-community structure and ecosystem processes, such as net primary productivity.

The decomposition study had three parts: a litterbag decomposition experiment in which litter from red pine (*Pinus resinosa*), red oak (*Quercus rubra* var. *borealis*), and red maple (*Acer rubrum*) was incubated in mesh bags in several places to detect the effect of the antenna after accounting for other factors that could also affect decomposition; an investigation of population densities of mycorrhizal-associated streptomycete populations; and an investigation of rates of disease progression through red pine plantations, the disease being infection with the root pathogen and wood-rotting fungus *Armillaria*.

Species and System Selection

Litter Decay

Red pine, red oak, and red maple are three dominant tree species of the northern hardwood forests of the Upper Peninsula of Michigan and in particular those of the antenna site. Litter from these three species represents the extremes (pine and maple) and the middle (oak) of a continuum of litter quality among species of the region, litter quality being defined as the sum of chemical properties of the material that affect decay processes. High-quality litter (maple) decays faster than low-quality litter (pine) because of higher initial nitrogen and water-soluble carbohydrate contents and lower lignin contents (Melillo et al. 1982). The choice of these species to represent the continuum of leaf-litter types in the area is justified.

Mycorrhizal-Associated Streptomycete Populations

Streptomycetes are common actinomycetes associated with myccorhizal symbioses; the exact nature of the association remains unclear (Marx 1982, but see Richter et al. 1989). Streptomycetes are important regulators of calcium oxalate, cellulose, and lignocellulose degradation (Crawford 1978, Knutson et al. 1980, Antai and Crawford 1981), perhaps through the effect of their excretion of vitamins, amino acids, hormones, and enzymes on heterotrophic decomposer populations (Richter et al. 1989). Streptomycete populations are good choices as indicators of microbial-community activity.

Root Pathogens

Armillaria is a heterotrophic fungus that decomposes lignin and cellulose in dead woody litter and live woody roots. *Armillaria* forms large, long-lived genets[1] (Smith et al. 1992) whose detection and culture are straightforward. *Armillaria* is a suitable organism for studying the effects of environmental changes on heterotrophic decomposer and disease organisms.

SELECTION OF RESPONSE VARIABLES

Litter Decomposition

Mass loss at various times during incubation was the only response variable selected. Mass loss integrates the activities of the entire microbial community as influenced by litter quality and extrinsic environmental variables, such as climate, soil water potential, and in this case the potential effects of increased electromagnetic radiation. It is unfortunate that analyses were not also made of the release rate of nutrients, particularly nitrogen, which do not necessarily follow mass loss in a straightforward manner. Mass loss is only a crude measure of the release rate of nutrients from decomposing litter for plant uptake (Melillo et al. 1982, 1984). The data required were in fact collected in the earlier years of the project but were later abandoned for logistic reasons.

Streptomycetes

The response variable chosen was the most probable numerical estimate of population density according to serial dilution in sterile cultures. This

---

[1] A genet is a single individual genetically identical with other individuals in a clone. For example, a single aspen tree is a genet arising from a root stock common to surrounding trees.

variable is the standard response variable in soil-microorganism ecological studies. Different morphotypes of streptomycetes were isolated by exposure to different stains.

Armillaria

Genets of *Armillaria* were isolated from different portions of the red pine plantations, and seedlings infected by each genet were mapped. The mortality of the red pine seedling population infected by a particular genet was the response variable.

EXPERIMENT DESIGN

Biologic and Ecological Sampling Techniques

The field design included the evaluation of the above response variables in two stand types (northern hardwood natural stand and red pine plantation) and three sites (overhead-antenna treatment, ground-antenna treatment, and control). A red pine plantation was sampled in each treatment site, and a northern hardwood stand was sampled in the overhead-antenna treatment and control sites. Thus, the experiment was pseudoreplicated (this will be discussed later). Fresh material for each year's placement of new litterbags was obtained from one nearby red pine and northern hardwood stand to minimize variation due to initial differences in materials. Litterbags were put into place each December and collected monthly in May-November the following year for 9 consecutive years; no individual litterbag experiment lasted longer than a year. Soil for isolation of streptomycetes was collected at the same monthly intervals as litterbags. Red pine seedlings were assessed annually for mortality associated with *Armillaria* genets.

Physical and Chemical Measurements

Intensities of the 76-Hz EMFs were mapped continuously at all sites, but the resulting data were never used in these experiments. The initial measurements used to determine treatment and control sites were the only measures of ELF-EMF exposure used.

Other climatic measures that could affect decay rates and microbial populations included annual sums of air and soil degree days, total precipitation and storm frequencies, and evapotranspiration, assuming a soil water-holding capacity of 25 mm.

Chemical measures of litter properties included initial concentrations of nitrogen, phosphorous, potassium, calcium, magnesium, and lignin in material collected each year and in monthly samples of incubated litter for the first 2 years only.

Statistical Methods

The main effect of interest—the treatment effect due to the presence of ELF EMFs generated by the transmitting facility—was pseudoreplicated (Hurlbert, 1984) in that there was only one site for each treatment. Therefore, the experimental data provide an estimate of variance of response variables within each treatment site but not the variance due to treatments across sites. Treatment is confounded with site. Sometimes pseudoreplication is necessary for logistical reasons, but apparently not in this case. However, when pseudoreplication is unavoidable, the generalization of treatment effects to other sites might be justified (with caution) if it can be demonstrated that the chosen sites for each single application of the treatment are not significantly different from each other at the outset and are at or very near the modal values of other environmental factors thought to affect the response variable of interest. Those two conditions were not considered, even though in many cases the data are available. The antennas were turned on in 1986; data on the physical and chemical measurements are available for 1985. For the litterbag experiments, there were statistically significant differences in annual mass loss across the sites in 1985 *even before the antenna was turned on* (Bruhn et al. 1994, pp. 78-80, Figs. 7-9). Thus, the treatment effect of interest is imposed on a pre-existing site effect. This pseudoreplication problem underlies many other studies in the overall ecological monitoring program. The acceptance of all further conclusions must proceed with this caveat in mind, but the caveat is not clearly stated anywhere in the report.

Because of covariation of site factors and treatments across sites, the effects of treatment, stand type, month, year, and (in the case of litter-mass loss) species were separated with an analysis of covariance (ANCOVA). Separate analyses were done for each response variable in each stand type; therefore, the overall effect in each stand type on the response variable of interest is unknown. Treatment (presence or absence of ELF EMFs) and year were combined into one variable, called "site-year." Month of sampling within a year was treated as a main effect. Replicates were three blocks or plots established within each stand at each site (the pseudoreplicate problem arises here) and replicates·of numbers of samples taken of litterbag, soil, etc. within each block.

For example, the effect of site-year and month on the percentage of original mass remaining at time t (X_t, $t = 1 \ldots 7$ months sampled per year) was tested with the analysis of covariance using nine climatic variables and one interaction term as covariates (Bruhn et al. 1994, p. 70, Table 23). The authors use the notation X_m for mass remaining at a given month m, but we will instead use the more standard notation X_t, where t is time measured in some meaningful units. The reasons for this will become apparent shortly.

This analysis of covariance is wrong, in that it treats elapsed time within a year (successive sampling months) as independent. For example, the decay in, say, July is not independent of the decay in June, May, or any previous month since $t = 0$, because the residual material left after decay in any month is passed along to the next month. Similar considerations apply to years. For example, although a new litterbag experiment was begun each year with material collected afresh from the reference collection stands each year, it is possible that the chemical quality of the material collected in year $t + 1$ is not independent of that collected in year t. Similarly, the populations of streptomycetes and especially the progression of *Armillaria* infection in year $t + 1$ certainly depends on the populations in year t, year $t - 1 \ldots$ year $t - n$. The data are therefore a time series, and the statistical tests used do not recognize this. The proper question to be asked is whether the presence or absence of the ELF EMFs alters the time series of data.

An additional problem with this analysis of covariance is that it examines the effects of month, year, site, and covariates on the mean mass remaining across all months within a given year. That mean is the average of all consecutive mass-loss data for months May . . . October. The mean of a monotonically decreasing function from $t = 0$ to $t = m$ months (see Eq. 1, below) is meaningless.

Sensitivity to and power to detect a statistically significant effect of the ELF EMFs on all response variables were considered, and the usual quality assurance and quality control procedures were used. However, the violations of some fundamental assumptions of ANOVA and ANCOVA and the pseudoreplicated nature of the design also affect the power tests. Therefore, any conclusions must be accepted cautiously, if at all, unless they are corroborated by alternative analyses.

Response to Review

The litter-decomposition and microflora study was peer-reviewed annually from 1983 to 1994. The comments of the reviewers seemed to be mixed. Suggestions that required minor changes in the experimental technique were usually followed, to the later satisfaction of the reviewers. More difficult problems involving theoretical issues of how to treat data were often not considered or only partially considered, even when the reviewer directed the researchers' attention to specific references in the literature to guide them. Very few of those references were cited in the final report, and this leads to the conclusion that the researchers did not understand the issues. For example, one reviewer repeatedly pointed out the utility of fitting decomposition models to a time series of mass-loss data and examining the effect of the antenna on values of the model parameters. The researchers responded by collecting such a time series of data but treating it improperly. Such problems remain in the final report and will be discussed below.

Presentation of Results

Consideration of Alternative Analyses

To critique the litterbag decomposition experiments, it is first necessary to consider some theory. Litterbag experiments generally rely on a standard first-order differential equation to describe decay rates (Olson 1963):

$$\frac{dX}{dt} = -kX, \qquad (3\text{-}1)$$

which hypothesizes that the loss rate of a homogeneous material X is a constant, k, under a given set of conditions that are themselves assumed to be time-invariant. The solution to this equation is:

$$X_t = X_0 e^{-kt}, \qquad (3\text{-}2)$$

where X_t is percentage of original mass remaining at time t. The authors chose not to pursue fitting the parameters of this equation to the data and evaluating changes in k as the response variable, because the first year's data showed that mass losses had positive residuals (discrepancies of the observed data from the predicted values) from this equation in early spring and late autumn and negative residuals in summer. That indicates that k is not constant with respect to time measured in months. In particular, the equation overestimates decay in spring and autumn and underestimates decay in midsummer; the magnitude of these residuals is not discussed. The authors conclude that the model is not a valid description of the decay process in this study.

They then postulate that might be due to changing environmental conditions during the decay process and therefore consider alternatives to include these environmental variables in Eq. 3-2:

$$X_t = X_0 e^{-k_1 t - k_2 p}, \qquad (3\text{-}3)$$

$$X_t = X_0 e^{-kpt}, \qquad (3\text{-}4)$$

and

$$X_t = X_0 (e^{-k_1 t} + e^{-k_2 p}), \qquad (3\text{-}5)$$

where p is an environmental variable, such as precipitation. They then reject those approaches because they would become "overly complex," particularly for the "mission oriented objective of this research" (Bruhn et al. 1994, p.15).

Nothing could be more wrong. First, what does it mean to say that Eq. 3-2 is the wrong model because of nonrandom distribution of residuals? That could happen for any of several reasons: (1) the model is fundamentally flawed as a description of the underlying biologic processes; (2) the units of the model, particularly those of the exponent, are not biologically reasonable; or (3) the model incorporates only some of the dynamics underlying the decay process, and others not modeled are responsible for the residuals. The authors accept reason 1 and do not consider reasons 2 and 3.

Let us consider the latter two reasons, whose elucidation is the normal manner of proceeding in studies of this type. First, although time in calendar units (movement of the earth in its orbit) is the usual metric for k, in fact such units have little biologic meaning for microorganisms except as they reflect the integration of some controlling factors. Rather, time in this case might be measured as the cumulative sum of some climatic parameter, such as actual evapotranspiration or degree days. When that is done, the problems of nonrandom distribution of residuals often disappear (McClaugherty et al.

1985). In fact, the presence of positive residuals under the cooler conditions of spring and fall and the negative residuals during the warmer summer supports such an approach. That was not considered.

Reason 3 is somewhat more serious, and in fact Eqs. 3-3 through 3-5 might represent attempts to resolve it. In fact, these equations are themselves the wrong approach. Rather, the more-appropriate model would be to replace k in Eq. 3-2 with a function describing a response surface of k with respect to various environmental variables:

$$X_t = X_0 e^{(f(\cdot)t)}, \tag{3-6}$$

where $f(\cdot)$ is a function describing a response surface of k in an n-dimensional space of environmental variables. Such an approach has been taken by Meentemeyer (1978), and Pastor and Post (1986). Although the strength of the ELF EMFs could be one of the variables, perhaps a more-powerful neutral-model approach could have been taken (Caswell 1976). In a neutral-model approach, the hypothesis would be that the response surface completely describes the behavior of k without any considerations of the ELF EMFs. Alternatively, it could be hypothesized that the presence of the ELF EMFs alters the shape of the response surface $f(\cdot)$. The advantages of such an approach would be that it flows directly from Eq. 3-2, which has real biologic meaning, and that it would be predictive. It is not at all clear that it would have been more complicated than the approach taken. The ANCOVA approach taken is not comparable with any approach in the literature, nor is it predictive. It is an open question whether it is simpler.

Interpretation

The major conclusions reached through the ANCOVA of litter-mass loss, streptomycete populations, and *Armillaria* infection rates of red pine seedlings are as follows:

- Very small changes in decay rates of each of the three species were statistically detectable between antenna and control sites; these differences were not consistent from year to year.
- No differences in streptomycete population of any morphotype between antenna and control sites were detectable.

- No differences in rate of spread of *Armillaria* infections of red pine seedlings were detectable, mainly because of high spatial variability in the spread of the disease.

The question is, given the above-noted problems of experimental design, Are the conclusions presented by the researchers warranted? Would alternative analyses support the conclusions (see Chapter 5)?

Litter Decomposition   The authors are correct in pointing out that a nonrandom distribution of residuals in the fits of data to Eq. 3-2 requires caution in the interpretation of the parameters of that equation. Nonetheless, k remains a useful first approximation of decay rates if the overall fit (r^2) is high and if the distribution of residuals follows the same pattern in all cases. The interpretation of differences in k would not then be biased by different and nonrandom distributions of residuals. We fitted Eq. 3-2 parameters to the mass-loss data for December 1, 1992 to October 31, 1993, as presented in Tables 3-5 (Bruhn et al. 1994, pp. 44-46) to see whether this alternative method of analysis corroborates their conclusions. December 1 was assumed as $t = 0$ in the absence of specific information in the report. Because small changes in k can have large consequences for mass loss, we also calculated half-lives (in days) of each species' litter for each treatment site:

$$t_{0.5} = \frac{\ln(0.5)}{k} = \frac{-0.69315}{k}. \tag{3-7}$$

The parameter values are presented in Table 3-1.

First, we consider the proportion of variation explained by Eq. 3-2 (r^2) and the distribution of residuals. Because $r^2 \geq 0.949$, the unexplained variation was always less than 5.1% and usually less than 2%. It seems that the exponential-decay model accounts for virtually all the variation across months for each species within a site. Therefore, the unexplained residuals, although admittedly not randomly distributed, are not serious. There is a pattern of slightly positive residuals during cooler weather in spring (the model overpredicts decay) that decline as the growing season warms (the model underpredicts decay), and occasionally return toward positive in fall in both the control and antenna sites. The pattern was common across all species and sites. Biologic reasons for this distribution of residuals have already been noted. Given the small residuals from Eq. 3-2 and their generally similar pattern, the use of k as a first approximation of decay rate poses no important problem. In

TABLE 3-1 Parameters Determined with Equation 3-2 for 1992-1993 Litter-Mass Loss

Treatment	k	r^2	1/2 life, days
RMAP	0.001713	0.996	405
RMAH	0.001619	0.996	428
RMCP	0.001630	0.995	425
RMCH	0.001476	0.992	469
RMGP	0.001675	0.999	413
ROAP	0.000957	0.964	724
ROAH	0.000923	0.954	751
ROCP	0.000884	0.953	784
ROCH	0.000883	0.949	785
ROGP	0.000879	0.965	789
RPAP	0.000864	0.985	802
RPAH	0.000951	0.977	729
RPCP	0.000963	0.990	720
RPCH	0.000905	0.986	766
RPGP	0.000903	0.988	768

RM, red maple.
RO, red oak.
RP, red pine.
AP, antenna plantation.
AH, antenna hardwood.
CP, control plantation.
CH, control hardwood.
GP, ground plantation.

fact, the use of k implicitly treats the data as a time series and avoids the problems in this regard noted for the ANCOVA.

Because of pseudoreplication, it is difficult to test for statistically significant differences in k between sites. This can be done by testing the difference in slopes of Eq. 3-1 in each treatment against that of the control; it is not attempted in this report. Nonetheless, an examination of the values shows that species differences account for most of the variation and that differences within a species between sites are relatively small. Moreover, the direction of difference between control and antenna sites is inconsistent between species. That is, sometimes litter decays faster in the antenna sites than in the control sites and sometimes slower.

Thus, the conclusions of the authors that the effect of the antenna on decay rates is much smaller than species differences and moreover is inconsistently expressed are borne out by the alternative analysis. The reason for the length of this alternative presentation is to show that such an alternative analysis, rejected by the authors, is possible and even results in simpler models than the ANCOVA used.

Streptomycetes   Insufficient data are presented to attempt a different analysis with, for example, a repeated-measures ANOVA or similar ANCOVA. However, inspection of the mean levels of micorrhizoplane streptomycetes and their coefficients of variation (CVs) for each month and each year (Bruhn et al. 1994, pp. 93-95, Tables 32-34) indicates that the means are rarely different in an absolute sense and the CVs are often in the range of 30-40% and sometimes higher. The adjusted mean levels of micorrhizoplane streptomycetes (adjusted for covariance of climate variables between sites) are different in each of the study sites only in the third significant figure. Given such small differences in the absolute values of the means and the large variation within a sampling unit, it is doubtful whether any other analysis could detect a statistically significant treatment effect. Therefore, despite the problems with the experimental design and ANCOVA noted above, the conclusions might be robust to the relaxation of assumptions.

Armillaria   Arguments similar to those for levels of mycorrhizoplane streptomycetes could be made for *Armillaria* infection rates. Inspection of the data in Table 42 of Bruhn et al. (1994, p. 116) for the percentage of red pine seedling mortality caused by *Armillaria* in the years 1986-1993 shows high variability with each site and similar means. In fact, the range of data for the control stand encompasses the range of data on the overhead-antenna site for every year and nearly encompasses the range of data on the ground-antenna site for every year. Given that distribution of data, it is doubtful whether any other analysis could detect a statistically significant treatment effect. Therefore, despite the problems with the experiment design and ANCOVA noted above, the conclusions are probably robust to the relaxation of assumptions.

CONCLUSIONS

The validity of these experiments for policy-making rests on whether the user of the results wishes to accept the problems involved in inferring a general effect of the antenna from a pseudoreplicated design and a design in which the progression of response variables is not treated as a time series of data. The user would have to accept the following propositions:

- That the effects or lack thereof of the antenna detected in the three sites is not due to other site differences that existed before and during antenna operation and that these effects or lack thereof will be present on any other set of sites under the influence of the ELF EMF generated (the pseudoreplication problem).
- That the projection of effects into the future does not depend on the past state of the system (the time-series problem).

The danger is in committing a type II error—accepting the null hypothesis (no effect) when it is false. It is not possible to calculate the probability of a type II error, because it depends on an independent estimate of differences between treatment and control. Inasmuch as the treatment effect is confounded with the site effect, the differences between treatment and control cannot be attributed solely to the antenna, the differences are not independent of pre-existing effects and confounding site effects that continued during the experiment.

Given the small effects detected and the large variance within sites and between years, the effect of the antenna, if any, on decomposition processes and the microbial community is most likely much smaller than existing natural spatial and temporal variation in the forests. We will never know, however, whether the results are site-specific. It should also be noted that this provisional conclusion of negligible effects of the antenna is due to the small differences in means and the large variation in the data and not to the statistical, mathematical, and biologic validity of the analytic methods used.

UPLAND FLORA

PROJECT PROPOSAL

According to Mroz et al. (1994), several studies (Wiewiorka 1990, Wiewiorka and Sarosiek 1987, Krizaj and Valencic 1989), which were independent of the Navy's ELF monitoring program, reported relationships different species between plant growth and EMF exposures. Accordingly, the effect of 76-Hz EMFs from overhead and ground antennas on tree growth, plant phenology, mycorrhizal symbioses, litterfall, and various environmental factors that could affect growth were assessed from 1985 to 1993.

SPECIES OR SYSTEM SELECTION

Red pine (*Pinus resinosa*) plantations and northern hardwood stands were chosen for study. These forest cover types are the most common in the Upper Peninsula of Michigan and are therefore appropriate for determining potential effects of ELF EMFs on tree growth and associated environmental variables.

SELECTION OF RESPONSE VARIABLES

Atmospheric, soil, tree-growth, litterfall, mycorhizal, and phenologic response variables chosen for study are listed in Table 3-2 with sampling frequency. This is an extensive survey of tree-growth measurements and environmental factors that potentially affect it. The sampling frequency for each variable is appropriate and in many instances greater than normally used in ecological research.

EXPERIMENT DESIGN

Biologic and Ecological Sampling Techniques

The field design included the evaluation of the above response variables in two stand types (northern hardwood natural stand and red pine plantation) and three sites (overhead antenna, ground antenna, and control). A red pine plantation was sampled in each treatment site, and a northern hardwood stand was sampled in the overhead and control sites. Replicates were three plots established within each stand at each site. Thus, the experiment is pseudoreplicated; this will be discussed later.

Physical and Chemical Measurements

These are included in Table 3-2. Researchers assigned a 76-Hz magnetic-field dose to each tree on the basis of a weighted-average dose from

TABLE 3-2 Variables and Their Sampling Intervals Tested for Significant Effects of 76-Hz EMFs in Upland Flora Studies (Mroz et al. 1994)

Atmospheric	Air temperature 2 m above ground (30 min) Photosynthetically active radiation (30 min) Relative humidity (30 min) Precipitation (cumulative, according to rain gauges)
Soil	Soil temperature 5 cm and 10 cm below surface (30 min) Soil moisture 5 cm and 10 cm below surface (3 h) Nutrients (0-15 cm mineral soil depth; total N and P; exchangeable Ca, K, and Mg; June and July)
Tree Growth	Diameter growth in hardwood species (1 wk) Height growth in red pine (1 wk)
Litterfall	Total mass by species (monthly or weekly, composited seasonally) Nutrient contents (N, P, Ca, K, and Mg; sampled as above)
Mycorrhizae	Total numbers by morphotype on red pine roots (monthly during growing season)
Phenology	Growth rates and timing of leaf and bud break on *Trientalis borealis* (twice a week)

all the times the antennas were on or off. However, the actual dose to each tree was never the average, because of temporal variation during the growing season due to the antennas being on or off and spatial variation of the field across each site. These problems are discussed further in Chapter 4 in the section on use of exposure data by ecological monitoring teams.

Statistical Methods

The main effect of interest—the effect of the ELF EMFs generated by the antennas—was pseudoreplicated (Hurlbert 1984), in that there was only one site for each treatment. Therefore, the experimental data provide an estimate of variance of response variables within each treatment at one site, but not the variance due to treatments across sites. Treatment is confounded with site. Sometimes pseudoreplication is necessary for logistical reasons, although that does not necessarily appear to be the case here. When pseudoreplication is unavoidable, the generalization of treatment effects to other sites might be justified (with caution) if it can be demonstrated that the chosen sites for each application of the treatment are not substantially different from each other at the outset.

Accordingly, an extensive survey of a number of candidate sites was made at the outset. Sites were compared for diameter and species distribution of trees and soil characteristics listed in Table 3-2, and other soil characteristics, such as texture and volume of rock fragments that could affect tree growth and species composition. Three sites were chosen to minimize differences in these variables at the outset to increase the probability of detecting an effect of the ELF EMFs. Analysis of variance (ANOVA) showed no statistically significant differences between sites in any property except diameter distributions of red maple, which had a greater proportion of large-diameter trees at the overhead-antenna treatment site. For logistic reasons, climatic and soil temperatures were not measured before site selection. However, a great effort was made to minimize site differences before antenna operation.

The careful site selection alleviated, but did not eliminate, the problem of pseudoreplication. The responses to the treatment were not evaluated within the context of variability across other sites that might be classified similarly.

More important, the conclusions apply only to these cover types on these soils. Although the cover types and soils chosen are common in the area, other cover types (jack pine, hemlock, and balsam fir) and soil types (hapludalfs[2] common on moraines) were not sampled. Therefore, extension of the conclusions to other upland ecosystems must be cautious.

To separate the effects of the ELF EMFs from other environmental variables, the data were treated with ANCOVA with repeated measures. Variables that had no demonstrated correlation with the strength of the magnetic-flux density of the ELF EMFs but did correlate significantly with response variables of interest were selected as covariates. For most response

---

[2]Hapludalfs are soils of temperate climates with an accumulation of clay in the lower layers.

variables (soil nutrients, litterfall, and plant growth), ambient climatic variables were used as covariates. The use of repeated-measures treatment of time is appropriate because the response in any year is not independent of responses in previous years. In the ANCOVA, the treatment-by-time interaction was examined for statistical significance. That is appropriate because it tests whether there are consistent differences in the effect of the ELF EMFs on the time series of data compared with controls. Because it tests differences in the shape and direction of the treatment responses, compared with controls, the treatment-by-time interaction term is independent of any site differences that existed before exposure to the ELF EMFs. The use of the treatment-by-time interaction term thus increases the power to detect statistically significant effects. Additional calculations of power and detection limits were also considered for each variable.

Tree-diameter growth was analyzed with a different method. The allometry (geometric relationship of the dimensions of one part of an organism to another) of diameter growth necessarily results in smaller increments in diameter with time regardless of limitations. The problem is to detect an effect of the ELF EMFs on the time course of diameter growth that is independent of other limiting factors. Accordingly, a model of diameter growth of hardwoods or shoot growth of red pine was constructed to predict expected growth given the selected tree species, size, site factors (mainly climatic), and presence of competitors. Residuals between predicted and observed growth were then tested for effects of the ELF EMFs by using the average strength of the field during the growing season (see comments above under "Physical and Chemical Measurements"). Parameter values for the model were fitted to local data and tested against tree growth observed before exposure. There were no consistent differences between predicted and observed diameter growth across all plots and in 2 years for red oak (*Quercus rubra* var. *borealis*), paper birch (*Betula papyrifera*), red maple (*Acer rubrum*), and quaking aspen *(Populus tremuloides)* or shoot growth of red pine. Details of model equations, parameter values, tests, and residual-effects testing are given in Mroz et al. (1994), Jones et al. (1991), and Reed et al. (1992, 1993).

RESPONSE TO REVIEW

The upland flora study underwent outside peer review for the years 1983-1994. In general, the reviewers praised the strengths of the study and the ability and willingness of the researchers to deal with the very difficult questions of experiment design. The use of the same reviewers during many consecutive years allowed a degree of continuity and accountability for making improvements in experiment design or improving explanations of what was being done. The reviewers seemed on the whole to be satisfied that their earlier comments were being taken into account or that their concerns were being addressed in later elaborations in the annual reports. For example, one reviewer seriously questioned the utility of the tree-modeling approach in the early years of the study but later explicitly reversed himself and agreed with the approach taken when it was more fully explained. That these researchers appeared to take peer review seriously might have contributed to their later success in publishing the results in peer-reviewed journals.

Presentation of Results

Consideration of Alternative Hypotheses

Other factors that might explain differences in response variables between control and treatment sites (climate, soil nutrients, etc.) were included in the design by virtue of the ANCOVA with repeated measures. One could argue at length about whether the methods of measurement of these covariates were proper or whether other variables should also have been considered. The fact remains that a substantial amount of effort was put into measuring a broad suite of factors likely to influence growth other than the ELF EMFs and establishing that there were negligible difference between sites in many of the response variables that were of interest before exposure.

Interpretation

The only variables that responded statistically significantly to antenna operation after consideration of covariate effects were in response to magnetic flux densities of 104 mG, with declining responses at higher fluxes:

- Soil temperature at a depth of 10 cm in hardwood stands (control 0.5-1.5°C warmer than overhead-antenna treatment site during the first 3 years of antenna operation, with declining differences thereafter)
- Increased diameter growth in aspen and red maple (about 0.14 cm/year maximal diameter response in aspen and 0.08 cm/year in red maple).
- Increased shoot (height) growth in red pine (maximal annual re-

sponses of 0.73 cm/year at the antenna-ground treatment site and 0.83 cm/year at the overhead-antenna treatment site).

Any changes in other variables during the period of antenna operation were explained by concurrent changes in covariates (particularly climate variables, but also variables related to stand maturation), rather than any residual correlation with ELF-EMF strength.

Conclusions

Validity

Among all the response variables measured (see Table 3-2 of this report), only four variables showed any response for which possible effects of the ELF EMFs could not be disregarded (soil temperature at a depth of 10 cm in the hardwood stand, diameter growth in aspen and red maple, and height growth in red pine at a narrow window of relatively low magnetic-field strengths). The responses of these variables are small and in fact did not persist as the magnetic-field strength increased. That is consistent with more-controlled experiments previously reported (Wiewiorka 1990, Wiewiorka and Sarosiek 1987, Krizaj and Valencic 1989). No detrimental effects on plant growth, phenology, or nutrient uptake were noted; if anything, there was slight stimulation of growth of the three tree species within a narrow range of small magnetic-field strength. Given the few, small, and ephemeral effects observed, the researchers concluded that the ELF EMFs did not have any statistically significant effect on vegetation and stand micrometeorology.

That conclusion needs to be tempered by the fact that the magnetic-field strengths used to assess exposure—the weighted average over the growing season—were not the magnetic-field strengths to which the trees actually were exposed. In fact, the trees almost always were exposed to a magnetic field either higher or lower than the average. In most cases, the magnetic-field strength is close to one of two levels. Thus, average strength is not likely to correspond to an actual exposure level. Therefore, the conclusion that tree growth is not affected, or even slightly stimulated, by exposure to a 76-Hz magnetic-field intensity of 2.0 mG is weakened by the lack of exposure to such a field strength. It can be concluded that the total operation of the antennas had no major detrimental effect on upland flora beyond the background environmental fluctuations of weather and other variables.

Uncertainties

The major uncertainty in the generalization of results lies in the pseudoreplicated nature of the experiment design. In part, that is an almost inevitable consequence of the site-selection criteria imposed to minimize site differences in as many variables as possible before antenna operation. It was probably extremely difficult to find three such sites with minimal differences in soils and vegetation, and they were then each divided into three replicates (or pseudoreplicates). Finding nine such sites to achieve true replication would be impossible. The demonstration that the sites were not different in almost every variable (except number of large red maples) before antenna operation only enhanced the detection of ELF-EMF effects if they existed at all.

The tradeoff is that, strictly speaking, the conclusions pertain only to the sites where the data were gathered. With some caution, they might be extended to other sites that appear similar (northern hardwood stands and red pine plantations of these densities on sandy haplorthods[3] of mixed mineralogy and frigid temperature regimes). However, the reliability of extending the conclusions to other sites is limited. Because of the pseudoreplication, there is no information on what the responses would be on other, albeit similar, sites. There is no direct evidence to determine whether ELF-EMF effects would be detected under other conditions. However, given the few, small, and ephemeral effects noted and the lack of any known mechanisms for the ELF EMFs to affect community structure and stand micrometeorology, this does not appear likely.

AQUATIC ECOSYSTEMS

PROJECT PROPOSAL

Potential effects of ELF EMFs on a stream ecosystem were examined by contrasting a stream site under the Michigan antenna and a paired control site downstream. The study had three main parts: examining the response of periphyton, analyzing the response of aquatic insects and their processing of leaf litter in the stream, and analyzing possible changes in fish abundance and movement as a result of exposure to ELF EMFs.

---

[3]Haplorthods are soils of colder climates with accumulations of iron and aluminum in lower layers.

Species and System Selection

Periphyton or algal production is crucial to any stream and is a natural place to look for the effects of environmental perturbations. The researchers examined aggregate production values (in terms of chlorophyll-a, cell volume, and organic matter) and the density of two dominant diatoms.

Insects could be influenced by ELF EMFs directly or indirectly through effects on periphyton. Insects were sampled with baskets buried in the substrate that could then be removed and sorted. Insect activities could also be influenced, and these were examined in two studies. First, the movement rates of a dominant dragonfly naiad were quantified in the presence and absence of ELF-EMF exposure. Second, the colonization of experimentally deployed alder leafpacks by insects was quantified as a function of ELF-EMF exposure, with measurements of leaf decay (quantified as a decay constant).

The abundance, community composition, and movement of fish were examined in the control and treatment sites. Because fish behavior or physiology might be altered directly by ELF EMFs, it is prudent to look at their responses of all these kinds. The individual species selected for more-detailed studies were the dominant species and thus a likely choice.

Selection of Response Variables

Periphyton

Periphyton response was measured both through aggregate variables (such as total chlorophyll) and through the densities of particular diatom species. These are apt response variables, although some indication of effect might be lost because of the aggregation of species into composite variables (see Carpenter et al. 1993).

Insect Response

The response of insects was quantified at the level of functional groups, species abundance, community indexes, movement of dragonfly naiads, and colonization of leafpackets. The only questionable response variable concerns the heavy reliance on "functional groups," such as "collectors-gatherers"; previous studies of aquatic systems have shown that these aggregate variables conceal effects evident at the genus or species level (Carpenter et al. 1993).

However, because several analyses were also done at the species level, this criticism is not too serious. The interpretation of such community descriptors as diversity indexes and evenness indexes is obscure without dissection into species responses, but these community descriptors are not given much discussion in any event. Naiad movement was analyzed by regressing average distance moved against ELF-EMF exposures. Average distance moved is a generally poor summary statistic for dispersal rates or patterns, and a more sophisticated dispersal variable would have been preferable (e.g., mean squared distance, which would give long-range dispersal greater weight, or the decay constant in an exponential probability distribution).

Fish Abundance and Movement

The analysis of fish response was to a large extent comparable with the analyses of insect response, with attention to abundances of particular species, and community structure. However, in this case, there was no aggregation of variables, and all the dominant fish were treated species by species. In addition to counts of fish, the size structure and growth rates of fish were assessed; this enhances the value of the study because these are likely to be more sensitive indicators of any effect than are census data. Finally, the fish moving past the treatment site and the control site were counted to see whether ELF EMFs inhibited movement in any way. Because ELF EMFs could alter movement, this was an excellent target for research, although, as discussed later, the questionable independence of control versus treatment sites compromised the results.

EXPERIMENT DESIGN AND IMPLEMENTATION

Only two sites were used: the treatment site and a control site 8 km downstream from the treatment site. The lack of replication is unfortunate, but the placement of the control site *downstream* of the treatment site makes the results of this study difficult to interpret. Such a placement means that one has to assume that what goes in one place does not influence fish, insects, and algae 8 km downstream—an unjustifiable assumption. The researchers selected the two sites to match physical conditions. But if the two sites must be on the same stream, it would be far better to have the control site *upstream* so that "turning on the antenna" is less likely to influence the control sites directly simply through mass movement of water, nutrients, and organisms downstream from the treatment site.

The streams were well characterized physically and chemically. Measurements included temperature, radiation, discharge, pH, dissolved oxygen, alkalinity, phosphorus, inorganic nitrogen, organic nitrogen, chloride, and silicate. The treatment and control sites were extremely well matched with respect to these physical attributes.

Statistical methods were sound, and the researchers did a good job of testing whether the assumptions of before-and-after, control-and-impact (BACI) were met and when they were not met, adjusting the analysis accordingly.

PRESENTATION OF RESULTS

Alternative hypotheses to ELF-EMF effects were examined where the BACI analyses indicated a statistically significant effect. For instance, possible alternative explanations of increased periphyton production of chlorophyll-*a* associated with ELF-EMF exposure were explicitly explored. Commendably, the power of the statistical tests was aptly examined; it was pointed out that only large changes in the fish community or biomass data could have been detected, given the variance and sampling frequency representative of the fish studies.

CONCLUSIONS

Validity

The methods and analyses of this study were generally sound. However, three weaknesses detract from it: lack of replication (only one pair of sites was contrasted); placement of the control site downstream of the treatment site, which might be expected to show changes at the treatment site and thereby undermine such analyses as BACI; and the fact that many of the response variables were aggregate variables, which might be expected to be poor indicators of change (Carpenter et al. 1993).

Uncertainties

By having such well-matched sites, the researchers minimized uncertainties due to fundamental site differences that had nothing to do with ELF EMFs. Unfortunately, because they placed the control site downstream of the treatment site, many of the findings of no difference between control and treatment sites are suspect. The researchers recognize this; when discussing fish responses (Burton et al. 1994), they point out that "it is invalid to assume that the ELF antenna operations do not affect the assemblage at [the control site]." The same uncertainty extends to other aspects of the ecological community as well, such as the insects and periphyton, although the researchers do not address that possibility.

The validity of the studies for policy-making is severely weakened by the placement of the control site downstream of the treatment site, even though the statistics are quite well done. If one dismisses the downstream problem, the evidence points mainly to a positive effect of ELF-EMF exposure on periphyton productivity and to no other major effects. It appears clear that the movements of dragonfly larvae, the colonization of leaf litter, and the movement of fish are not inhibited in any way by ELF EMFs. The ability to detect changes in the fish community was weak, as mentioned by the researchers, so only a very large change in fish biomass or abundance could have been detected, given the research design and inherent variability in the data.

POLLINATING INSECTS

PROJECT PROPOSAL

The proposal is clearly stated and justified. If bees are disoriented or damaged by ELF-EMF exposure, orientation to nests and foraging behavior are likely to change. If foraging is slower and less efficient, bees might reduce reproductive investment per progeny. Progeny might have higher mortality because of both direct and indirect ELF-EMF effects.

To address the potential effects of ELF EMFs, the researchers generated eight specific, testable hypotheses. Possible mechanisms and response variables are discussed for each hypothesis. Other factors that could affect the responses are considered later in the discussion.

SPECIES SELECTION

The choice of two species of leaf-cutting bees (Megachilidae) is fairly well justified on mechanistic grounds. The orientation and behavior of honeybees have been shown to be affected by high-voltage transmission lines and fluctuations in the earth's magnetic field. Honeybees are known to have magnetite particles in an abdominal organ that they use as a compass. The weak point, which is noted, is that "no data exist on the ability of megachilids to detect magnetic fields" (Strickler and Scriber. 1994, p. 5) or on whether they also contain magnetite. And the differences between fields resulting from the ELF antenna, fields from high-voltage transmission lines, and the earth's magnetic field are not discussed.

Limiting the study to two bee species changed the research from the initial proposal to study a bee community and yet allowed for comparisons and reduced the risk inherent in studying one species in the field. Apparently, the two species were expected to respond similarly. Honeybees might have afforded a better biologic assay of effects because of previous studies, but honeybees are rare in the study area and cannot overwinter there. However, some behavior studies could have been done with imported hives.

The two megachilids were excellent bee species to study because they will use artificial trap nests placed out by the researchers. Therefore, nest availability and location could be controlled, and aspects of the nests and their contained progeny could be measured. Both species were relatively common in the area, so sample sizes were usually adequate.

SELECTION OF RESPONSE VARIABLES

The decision to examine behavior, reproduction, and mortality, rather than population sizes, was excellent, given that population estimates are notoriously inaccurate. Instead, good measurements of specific responses that could affect population growth were made. The response variables were mostly based on previous results of studies of honeybees exposed to high-voltage transmission lines and generally were highly relevant and well justified.

Honeybees show changes in orientation (waggle dances) and increased agitation in the presence of EMFs. For these bees, duration of leaf foraging was measured because it was more consistent than pollen and nectar foraging. Trip duration could be affected by such factors as leaf distance or quality, but these were not examined. Trip duration probably involves less-complex behavior than honeybee orientation and might be less likely to be disrupted by exposure to ELF EMFs. Orientation was measured as direction of nest choice.

Honeybees exposed to high-voltage power transmission lines produce fewer cells and collect more propolis (a resinous substance in tree buds). Nest-building behavior was characterized by cell length, cells per nest, plug thickness, and leaves per nest. However, all those factors and reproductive investment per offspring depend on the sex of each offspring and are autocorrelated. Mothers under stress can vary the sex of their offspring. Sex ratios could not always be directly determined because of deaths and parasitism; therefore, they were estimated on the basis of cell positions; this adds some error.

Overwintering mortality of honeybees has been shown to increase under high-voltage power transmission lines. Overwintering mortality was measured as the percentage of progeny dying and the percentage of nests with at least one death.

Although each of those responses could change with ELF-EMF effects, they can also change in response to other variables—such as resources, temperature, or bee age—and can be determined by a complex of factors that can vary spatially and temporally.

The researchers could not estimate total reproductive output (an index of fitness) because females can make more than one nest and individual females were not marked and followed.

EXPERIMENT DESIGN AND IMPLEMENTATION

Biologic and Ecological Sampling Techniques

Sites were chosen on the basis of required exposure criteria. However, the experimental design has two major flaws. First, the two replicates are actually pseudoreplicates because they are so close together; the researchers addressed this by using a nested design. Second, the treatment and control areas initially differed in many respects, including flower resources, and these probably changed independently in both time and space.

Otherwise, great care was taken in standardizing methods and measurements. This often involved complex procedures. The methods of sampling (nest boxes) measurement are discussed in detail and could be replicated by new researchers. Observer bias was included in the analysis. The study is rigorous.

The researchers provided 1,152 nests per site, ensuring nonlimiting availability and good samples. Completed nests were immediately replaced with new trap nests.

Overwintering mortality was measured on the basis of nests kept outside starting in 1989; they would not be affected by 60-Hz EMFs and microclimate indoors.

Physical Measurements

Climatic measurements were obtained from the ecological monitoring program's upland flora study. However, these were not used in the final analyses—a reasonable decision, given that local microclimate probably affects bee behavior more. Nest hutches were exposed to different temperatures and shading. Because of complex variation over time, measurements of this were not taken.

Statistical Methods

Data were analyzed by using the general linear models (GLM) procedure on a Statistical Analytical Systems (SAS) software package. Given that the design was unbalanced, with both nested and random effects, a mixed-model analysis of variance (ANOVA) using maximum likelihood or a before-and-after, control-and-impact (BACI) analysis would be more appropriate. Whether the final interpretations of results would be different is unclear, given the high variation, confounding variation, and lack of true replicates. When distributions were nonnormal (for nest plugs), the researchers appropriately used a nonparametric factorial analysis of the data.

The authors were aware of those statistical problems and addressed many of them. Their test for statistically significant ELF-EMF effects was based on a significant interaction between treatment-site response variables and levels of antenna operation, which would indicate that differences between the treatment and control sites were greater when the antenna was on full power than earlier, when it was on low or 50% power. However, this is problematic, as recognized, in that such an interaction could indicate either that the treatment and control sites respond differently to the different antenna levels or that the sites were different before antenna operation, but not after, for other reasons.

For each hypothesis, minimal detectable differences and power of the tests were calculated and discussed in relation to the results.

Quality Assurance and Quality Control

In general, great care was taken in measurement. For example, measurements of nests were made in wire-mesh Faraday cages to minimize exposure of developing bees to electric fields during measurements. For consistency, only the first three foraging trips were measured. Observer identity was used as a variable in the ANOVA model to test for bias. Because sex ratios can vary with nest diameters and depths, data not standardized for these variables were excluded from the analysis.

Exposure Assessment

Measurements of actual ELF-EMF exposures are reported for the treatment and control sites, with maximums and minimums given for one of the two treatment sites for June through August. The 76-Hz magnetic-flux densities at the treatment site were 330 times stronger than at the control site.

Exposure varied substantially between treatment sites, within treatment sites, and among years. The two treatment sites differed, in cumulative gauss-hours, by a factor of 2 (Strickler and Scriber 1994, Fig. 13). Nest hutches differed within a treatment site by up to a factor of 100. Pretreatment years included low to 50% power (1983-1988). Mobile adult bees are subject to different exposures, but developing progeny are stationary and should have a consistent exposure, although this obviously varied between nests, sites, and years.

The researchers argue that this variation in ELF EMFs is unlikely to be important because the bees will show a threshold response to the large exposure differences between treatment and control sites. However, if the responses are dose-dependent, this variation could confound the results. If there is a humped response curve (maximum at some intermediate level), ELF-EMF effects would be missed in the analysis.

RESPONSE TO REVIEW

Reviewers' suggestions to focus the initial study were heeded and contributed greatly to the success of this research. Other suggestions about methods, such as shielding overwintering progeny from electric fields indoors and including observer bias as a variable, were also followed.

Two major suggestions were not followed, for reasons that are not entirely clear. One was to use a BACI analysis with covariates. The other was to increase replication by placing fewer trap nests at many sites. The original design had four sites (with no replicates), but two of these were dropped. Given that ELF EMFs become ambient at 1.6 km, one reviewer suggested that control and treatment sites could be closer than 48 km apart, which could reduce some of the other environmental differences between sites. Increased replicates and reduced variation could have allowed for stronger conclusions about the results.

Presentation of Results

The authors are forthright about problems, such as unreliable data, which were eliminated from the analysis, and other confounding variables. A strength of this report is the clarity of discussion about problems and alternatives.

Alternative hypotheses were often used to dismiss results that indicated potential ELF-EMF effects. The researchers' arguments are reasonable and soundly based on the data, but opposing arguments could be made for real ELF-EMF effects in most cases. Of eight parameters analyzed for one or both species, five showed statistically significant ELF-EMF effects for one or the other species, but only one, overwintering mortality, is interpreted as a possible real ELF-EMF effect and this effect is regarded as ambiguous.

Statistically significant effects on three parameters (cell length, leaf number, and nest orientation) are interpreted as being caused by factors other than ELF EMFs. Cell lengths became more similar after full antenna operation for one species. The authors argue that ELF EMFs were not likely to have prevented the reduction in cell length (0.2 mm of 11.1 mm) at the treatment site. Leaf number also became more similar after the antenna began operating at full power for one species; that is the opposite of what is expected if ELF EMFs are detrimental. The authors also argue that this difference (0.5 leaf/cell) is trivial and unlikely to affect fitness or population growth. Nest orientation was argued as reflecting differences in local flower availability or shading, rather than ELF-EMF exposure.

Effects on nest thickness were not statistically significant, but the authors noted that the effect would have had to be very large to be detected with their test (hence, they dismissed this as a strong conclusion). They also cautioned against accepting the null hypothesis for trip duration and sex ratios, because of low power; that is commendable. In contrast, they accept the null hypothesis of no effects on offspring weight because statistical power was good.

The ELF-EMF effect on overwintering mortality was statistically significant for one species. Mortality was lower at the treatment site at low antenna power but increased to the level of the control site after full antenna operation.

The authors dismiss this as an ELF-EMF effect because mortality became increasingly similar and because of protocol changes, parasitism, and small sample sizes. However, mortality did increase after full antenna operation.

Overwintering mortality was also found to differ significantly with ELF-EMF exposure in a nest-transplant experiment. Nests occupied at a treatment site but moved and overwintered at control site showed lower mortality than nests occupied and overwintered at the treatment site. In addition, the mortality of transplanted nests was similar to that of the nests occupied and left at the control site. Although that is strong evidence for ELF-EMF effects, the authors argue that the effect is unlikely to be caused by ELF-EMF exposure.

The sources of mortality were not emphasized in the researchers' report, because the sources were difficult to determine. For example, the researchers could not separate prepupal mortality in winter from that in summer, fall, or spring. They speculate that weather probably was important for much of the prepupal mortality. Because prepupal mortality varied greatly between years and sites, they decided to test the potential effects of ELF-EMFs on mortality after a bee had survived to the prepupal stage. That would reduce the variation caused by site and weather differences. For this study, a major parasite was the cuckoo bee, *Coelioxys*, which is also a megachilid and could not be distinguished from the host in the prepupal stage. Therefore, both parasites and hosts are included in the mortality data. They argue that hosts and parasites would probably have been similarly affected by ELF EMFs so this should not distort the results or interpretations.

Overall, the authors conclude that ELF-EMF impacts are absent or at most minimal.

Conclusions

Validity

Two major problems in the design seriously reduce the ability to detect statistically significant ELF-EMF effects: low replication (pseudoreplication) and confounding variables. The treatment and control sites initially differed in many factors, including flower resources, which also probably varied independently over time.

Otherwise, the details of the study were well designed and well executed. Great care was taken in standardizing variables and in maintaining data quality. A more appropriate statistical analysis could have been used, but it is doubtful that different conclusions would be reached. Problems are discussed at length.

Uncertainties

The authors' conclusion that ELF-EMF effects are absent or minimal is uncertain because of the weak ability to detect effects. That results in a greater likelihood of accepting a false null hypothesis (type II error) than of rejecting a true null hypothesis (type I error or false positive). A type II error would also be likely if the response curves were dose-dependent and either monotonic or hump-shaped. Therefore, the fact that the authors did find statistically significant effects requires careful consideration. The finding of increased overwintering mortality in two independent experiments makes an especially strong case for the existence of statistically significant ELF-EMF effects.

The researchers' conclusion that ELF-EMF effects are absent or minimal might reflect the low power of the tests rather than the reality of no effects. Real effects would likely have been difficult to detect because of the small sample sizes and high variation in many factors. Therefore, the conclusion of "no effects" might, in fact, be based on the acceptance of a false null hypothesis (type II error). A type II error would also be likely if the response were dose-dependent and showed a monotonic or hump-shaped relationship to dose.

To their credit, they discuss their reasons at length, but good reasons could be advanced for rejecting the null hypothesis in most cases. They also argue that the few statistically significant effects are small and would have little impact on populations. That is an erroneous argument because small differences over a long time can produce large changes in population sizes.

Summary

The authors' final conclusion that ELF-EMF effects are absent or minimal is questionable. The authors' explanations are inadequate for discounting potential ELF-EMF effects and for accepting the null hypothesis of no effect. Given the weak ability of the experiment design to detect ELF-EMF effects, any significant effects should be given careful consideration. Similar arguments were made by one reviewer of the final report who strongly feels that ELF-EMF effects were clearly demonstrated in this study. More independent replicates would help to determine whether the statistically significant effects are caused by ELF EMFs or by some other factor.

SOIL ARTHROPODS AND EARTHWORMS

PROJECT PROPOSAL

Soil macrofauna, such as arthropods and earthworms, control many of the decomposition processes critical to ecosystem function. It makes sense to examine the abundances, activity, and demographics of dominant soil animals. This study has six parts: soil and litter arthropod censuses, analyses of surface-active arthropod activity via pitfall traps, analyses of earthworm populations sampled by square cores, analyses of growth and reproduction rates of earthworms incubated in soil bags, analyses of litter inputs sampled by litter traps, and analyses of litter-decomposition rates measured in litterbags.

SPECIES AND SYSTEM SELECTION

Soil and Litter Arthropods

The species chosen for analysis were essentially the numerically dominant mites and collembolans (springtails) found in soil cores and litter samples. With no reason to choose a particular species, it is sensible to use the dominant species, because their frequency of occurrence makes them conducive to statistical analysis, compared with organisms that occur in only a few samples.

Surface-Active Arthropods

The researchers selected species on the basis of their abundance and commonness, which is reasonable. It would also have been desirable to pick species on the basis of relative uniformity in distribution over the areas of interest, so as to diminish the statistical problems of place-to-place and time-to-time variations. In addition, arthropods that actively forage on the soil surface could exhibit sublethal effects because of environmental perturbations (through altered behavior), and their activity patterns are potentially good indicators of ELF-EMF effects.

Earthworm Field Populations

The researchers examined all nine earthworm species found at the treatment or control sites. Because they looked at *all* species, there was nothing arbitrary in the analysis, and the thoroughness is commendable.

Earthworm Growth and Reproduction in Incubation Bags

In addition to counting animals, it is a good idea to look for per capita differences in reproductive rates or growth. Using such demographic characteristics can yield far more sensitive indicators of ELF-EMF effects than waiting for population densities to reflect differences due to the activation of the antenna.

Litter Inputs

One can imagine ELF-EMFs influencing tree phenology, leaf production, and leaf abscission in a way that could alter litter inputs into forest soils. The measurement of litter inputs represents a sound research decision.

Litter Decomposition

Litter decay is certainly an appropriate system process to examine, especially in the context of concordant measures of earthworm and other macrofauna associated with decomposition.

SELECTION OF RESPONSE VARIABLES

Most of the response variables involved counts or densities of individuals by species, an unassailable focus for analysis. Measures of community structure, such as diversity and evenness indexes, were also examined; these aggregate indexes are difficult to interpret, and any changes in them would be impossible to understand without also analyzing effects species by species. However, these community indexes play a minor role in the analyses and are distracting only in that they add length to an already confusingly long document.

A few response variables deserve comment. For the earthworm component, the vertical distribution of worms was examined as a possible indicator of changes in behavior or habits. The age distribution of worms and the size and cocoon production of earthworms were examined as functions of year (preoperational versus operational antenna years). In the earthworm-incubation experiments, the growth and reproductive status of "enclosed" earthworms were followed as a function of antenna operation and previous exposure. Because earthworms can live several years and might require 3 or more years to reach maturity, the attention to growth and reproductive rates in the earthworm component of the study is commendable; these demographic rates should be more sensitive than absolute population numbers (which could reflect long time lags and history). Leaf litter inputs were assessed by lumping grams of dry weight per square meter for basswood, maple, and all other species. Litter decomposition was assessed on the basis of the percentage of initial mass remaining. Mass loss is a crude measure of decomposition, and it would have been far better if a more-direct measurement of nutrient release had been obtained.

Experiment Design

Biologic and Ecological Sampling Techniques

The field design involved a single control site and a single treatment site for all components of the research. Replication is therefore impossible, and the only appropriate statistical analysis uses a before-and-after, control-and-impact (BACI) approach. Although the absence of replication might be unavoidable, some aspects of the rationale are dubious. First, the two sites were not comparable: they had strikingly different earthworm and arthropod fauna. Second, for the earthworm-incubation experiments, the experiment design involved moving earthworms to a control site at which they did not occur naturally in any abundance (compared with either the treatment site or the sites from which they were taken). Thus, the experiment could be interpreted as an investigation of the effects on a species' growth and reproduction of transplanting it out of its habitat. The detailed studies of earthworm size, age, and reproductive structure were confined to species that occurred only at the treatment site. That makes it virtually impossible to distinguish possible effects of antenna activation from effects of other temporally varying factors.

Physical and Chemical Measurements

ELF EMFs were measured, and it was well demonstrated that the treatment site exhibited ELF-EMF levels at least 10 times those at the control site. The sites fulfilled IITRI's other criteria as well.

Statistical Methods

The studies had four severe statistical flaws. First, when the BACI analysis was used, the appropriateness of its assumptions of no serial autocorrelation and of additivity was never evaluated. Because the control and treatment sites clearly differed before the antenna was turned on, this is a serious problem. Second, often a simple ANCOVA was applied, although a repeated-measures approach is more appropriate. Third, the power of the tests and sampling scheme was low. For example, to sample surface-active arthropods, only 10 pitfall traps were used per site, an absurdly low number. In addition, given the large differences between sites and the variability in data, one questions how powerful the BACI could be in detecting ELF-EMF effects. Fourth, for many of the earthworm analyses, the only data came from the test site, and no serious effort was made statistically to distinguish temporal variation due to antenna activation from temporal variation due to other environmental variables.

PRESENTATION OF RESULTS

Consideration of Alternative Analyses

In general, analyses did not include much consideration of alternative approaches. For example, numerous BACI tests were performed, some of which indicated statistically significant effects. No serious effort was made to determine whether those effects were due to ELF EMFs or to some other environmental characteristics associated with the difference between control and treatment sites.

Interpretation

The interpretations of data seem predisposed to conclude that ELF-EMF exposure had no statistically significant effect; several potentially significant results (as detected via BACI) were consistently dismissed. Biologically, that might be the correct conclusion, but the interpretation gave too little attention to the weakness of the experimental approach and the statistics being used relative to the variance in the data. Some experiments, such as the earthworm-incubation studies, were especially questionable in interpretation. This study cannot distinguish between a test of "natural" versus "unnatural habitat" and treatment versus control, because in the control site the earthworm species being incubated was very rare.

Conclusions

The utility of this research for policy-making is compromised by a failure to ensure that BACI was properly applied, by the problem of pseudoreplication and the fact that at best only one control site was compared with one treatment site (which differed substantially from the control site before antenna activation), and by a statistical failure to disentangle temporal trends due to antenna activation from other temporally varying environmental factors. Any effects of ELF EMFs were small, compared with the total variation of the processes measured over the 11-year study; but the study included drought years (likely to be important to soil macrofauna), and it might not be consoling to learn that variation due to ELF EMFs is minor in comparison with the variation caused by a severe drought.

SOIL AMEBAS

Project Proposal

This study, carried out over a period of more than 10 years starting in 1983, records data on populations of soil amebas from two treatment sites near the antenna (one next to the ground terminal and one under the antenna) and a control site about 15 km away, where 76-Hz EMFs were at most one-tenth as large. The studies focused on the ameba *Acanthamoeba polyphaga* and included counts of organisms in the soil (population sizes), growth-rate measurements in situ in culture vessels designed to match ELF-EMF exposures in the soil, and species present and determinations of genetic heterogeneity based on isoenzyme analyses.

Species Selection

Acanthamoeba polyphaga is one of the more common and already well-studied species of soil amebas that occur near the Michigan transmitting facility. Soil amebas are micropredators, and variations in total counts are thought to reflect differences in quality and quantity of food available, especially bacteria. They are close to the bottom of the food chain and thus might be highly indicative of effects at that level. Measurements of total bacteria in the soil, a classically indeterminate value, were attempted with a modification of the acridine orange direct-counting technique; numbers were around 109/g of soil but were highly variable, so attempts to use this technique to explain variance in soil-ameba numbers were abandoned.

Selection of Response Variables

The three principal response variables, as noted above, were counts of organisms in the soil, growth-rate measurements, and determinations of species present and the genetic heterogeneity. These seem well suited to address the overall study from an ecological vantage point, in that they embrace the static situation, the dynamic growth question, and possible genetic effects according to a well-established criterion.

Experiment Design and Implementation

Biologic Sampling Techniques

The sampling techniques were well described and appear to have been carried out with professional competence. Similar experimental procedures were used for samples from the control and treatment sites.

Physical Measurements and Sites

The physical measurements of the fields and currents at both treatment and control sites were made in cooperation with IITRI personnel and are well described in the report. Extensive measurements were also made of soil chemistry and moisture, the latter having a substantial effect on the biologic systems measured. Data on temperature were collected and are presented in full.

The sites for sampling the organisms were selected in cooperation with IITRI personnel. All sites had a similar 60-Hz EMF background, and the control site had 76-Hz EMF intensities no more than one-tenth those at the treatment sites. The ground treatment site was 39 m from the ground terminal, whereas the antenna treatment site was about 40 m from the north-south leg of the antenna. The control site was 15 km south of the ground treatment site.

Statistical Methods

One-way analysis of variance (ANOVA) was used to detect differences in total-ameba and cyst counts at the three sites. The before-and-after, control-and-impact (BACI) analysis was used for the log maximal ameba counts and the maximal cyst counts and in the measurements of genetic diversity.

CONCLUSIONS

No population-density differences were found between antenna treatment, ground treatment, and control sites. In addition, growth rates of *Acanthamoeba polyphaga* did not differ between sites, and genetic-diversity studies failed to reveal differences between sites. The only statistically significant difference found was in conditions before and after antenna operation; there was a small but statistically significant difference in maximal population densities between the control site and ground treatment site, but by the same method of analysis no differences between the control site and the antenna treatment site or between the antenna and ground treatment sites.

The study appears to have been carried out carefully and competently. The inherent variability in the data was great and changes due to temperature and moisture were large, so small effects would not have been detectable. But the study results constitute convincing evidence that large-scale ELF effects did not occur over the time that the study was carried out.

It should be noted that the antenna was fully operational only during the last years of the study and that even then there were down periods. Nevertheless, this study was adequate for the questions being asked.

4

Common Issues

THE COMMITTEE'S EVALUATION of the 11 ecological studies that were included in the Navy's ELF ecological monitoring program revealed several issues that were common to many or all of the studies. Those common issues are discussed here. This chapter was not intended to discuss each study in the context of each common issue. Specific studies are discussed below for illustrative purposes. Overall conclusions and recommendations concerning the common issues are presented in Chapter 5.

USE OF EXPOSURE DATA BY ECOLOGICAL MONITORING TEAMS

All ecological monitoring teams made use of the division between treatment and control sites. All pairs of sites satisfied the criteria for 76-Hz exposure except the aquatic study sites, for which the ratio of electric fields in the earth at treatment sites to those at the control sites did not quite satisfy the criteria (see Chapter 2). That was rectified by inclusion of supplementary sites closer to the antenna in 1990. About 10% of the study sites did not satisfy the criterion for 60-Hz exposure, but in all cases the 60-Hz exposures were very low. From an ELF-EMF exposure point of view, the pairing of sites into treatment and control was satisfactory. IITRI was successful in characterizing the ELF EMFs for this purpose.

In some studies, it was important to know whether the transmitter was on or off during critical "exposure" periods. Those were the studies that used response variables potentially sensitive to ELF-EMF exposure for only short periods. For example, three of the four major completed studies of small vertebrates considered short-term phenomena: embryonic development (4 days or less), homing (4-5 hours), and maximal metabolic rate (minutes to perhaps 24 hours). The fourth study, assessing fecundity, took place over a period of weeks to 2 months, depending on the year; the investigators' division into treatment and control sites might not have corresponded to the actual exposures, which were most likely time averages of antenna activity over that period.

Without verifying that the transmitter was on during the experiments, it is not possible to know whether a treatment site was exposed to ELF-EMFs from the antenna. And without that knowledge, there could be misclassification of subjects from the control category into the treatment category. However, there is little evidence that most of the investigators were aware of or considered this factor in evaluating the results of their experiments. Chapter 5 discusses possible reanalysis of the data collected through some studies.

According to the monitoring-program reports reviewed by the committee, only the upland-flora team tried to use any of the other data provided by IITRI, although some teams requested additional ELF-EMF data. Investigators from the earthworm and soil arthropods study requested and received information on electric field vs. soil depth but never used it. Extensive measurements of the electric field in the earth were made at the Martel's Lake site (overhead antenna treatment site for upland-flora and litter decomposition and microflora studies) but never used. The authors of the pollinating insects study indicated that each hutch received a different exposure, but there is no indication of an attempt to use these data in the analysis. Some of the teams might have decided that there was no difference in ecological aspects between treatment and control sites and hence that there was no need for further analysis with specific ELF-EMF exposure data. Sometimes, however, only the consideration of additional exposure hypotheses will identify an effect. Such extensive exposure data were available, and they should have been used. The use of ELF-EMF exposure data in two specific studies is discussed here.

ELF-EMF CHARACTERIZATIONS AT WETLANDS SITE

In 1986 and 1987, the wetlands researchers conducted a series of stomatal-resistance measurements on wetlands plants. The measurements were related by multiple-regression analysis to the independent variables: two environmental variables and the magnetic field and electric field in the earth resulting from the antenna system under full-power conditions as measured by IITRI. The purpose of this analysis was to determine whether there was a statistically significant relationship between stomatal resistance and any or all of the independent variables.

The researchers noted that the antenna on or off condition was neither predictable nor observable. They correctly noted that this was a potentially confounding factor in the analysis of short-term responses because measurements might have been taken at a treatment site when ELF-EMF exposure was not occurring. They did not pursue the analysis, because they found inconsistent results under a variety of both similar and substantially different exposure conditions.

USE OF MAGNETIC-FIELD INTENSITY AS A "DOSE" AT UPLAND-FLORA SITE

In the upland-flora study, the researchers designated a 76-Hz magnetic field as an indicator of dose to each tree. According to one of their papers (Reed et al. 1993), the indicator was based on "average exposure to magnetic flux density during that particular growing season." It does not appear that the actual measure used in the study is consistent with this definition.

The measured fields across the treatment area (during transmitter operation at full power) varied from about 5-10 mG (see Haradem et al. 1994, p. D-7, for location of points 4T2-6,7,8,12,13,26,34 on the hardwood stand and p. D-30 for the historical measurements at these points). However, the upland-flora report shows magnetic-field measurements of about 1-9 mG and a specific "effect" on tree growth at about 2 mG. The researchers would have been expected to define the growing period and use antenna on and off time to derive an "average" field during this time. According to the researchers, the growing period for hardwood trees was about April through September. The antenna on-time statistics for each antenna configuration allow calculation of the average magnetic field over the growing period. These are shown in Table 4-1; the statistics behind the calculations can be found in Appendix J of the final engineering report (Haradem et al. 1994).

It is clear from Table 4-1 that no trees were exposed to magnetic fields within the range of 0.3 and 3.0 mG. Thus, the method described here is not

TABLE 4-1 Range of Average Magnetic Fields During the Growing Period Over the Martels Lake (Overhead Antenna) Treatment Hardwood Stand by Year

Year	Effective Magnetic Field as a Percentage of Full-Power Magnetic Field	Range of Magnetic Field for Any Tree within the Treatment Site, mG
1986	0.05	0.0025-0.005
1987	0.3	0.015-0.03
1988	2.1	0.1-0.21
1989	59.0	3.0-5.9
1990	92.0	4.6-9.2
1991	63.0	3.2-6.3
1992	84.0	4.2-8.4
1993	93.0	4.7-9.3

the method actually used by the authors to determine the exposure. Instead, the authors used spot measurements made during operation of the transmitter, which were provided by IITRI. Consider the points 4T2-6 (closest to the antenna) and 4T2-7 (farthest from the antenna) on the hardwood-stand treatment site. The measurements provided by IITRI and the number of hours that each point was exposed at the measured levels during the growing season are shown in Table 4-2, derived from tables in the final engineering report (Haradem et al. 1994).

Note that exposures that correspond to field measurements between about 1.2 and 2.6 mG are within the range shown in Table 4-2 only for 1991. That is important because the claimed effect occurs at about 2.0 mG. Furthermore, during that year, the exposure was more often in the range of 5.4-10.3 mG than it was in the range of 1.6-3.0 mG.

The following observations can be made about this study. The indicator of dose reported in publication of the upland-flora work was not that actually used by the authors. They reported that the indicator of dose was defined as "average exposure to magnetic flux density during that particular growing season." The indicator actually used was based on spot measurements while at least one antenna was on. The authors did not provide a clear rationale for the indicator of dose that they used. Specifically, they were not clear about

TABLE 4-2 Spot Measurements of Magnetic Field at Hardwood-Stand Treatment Site and Number of Growing-Season Hours at These Levels

Year	Measured Magnetic Field at Point 4T2-6, mG	Measured Magnetic Field at Point 4T2-7, mG	Total Time Exposed During Growing Season (4,392 Total Possible Hours), h
1986	0.73	0.37	24.8
1987	1.16	0.59	142.8
1988	5.0	2.6	162.2
1989	10.3	5.4	2,390
1990	11.0	5.8	3,795
1991	3.0	1.6	1,462
1992	10.3	5.4	3,701
1993	10.3	5.5	4,073

Notes: In 1986, there were 24.8 h of operation at levels indicated and 17.0 h of operation in which measured levels were 0.44 and 0.22 mG. It is not clear why values in the table were used, rather than those reported here or zero, which was by far the most-common exposure level during growing season.

In 1991, there were 1,462 h of operation at levels indicated and 2,336 h of operation in which measured levels were 10.3 and 5.4 mG. It is not clear why values in the table were used, inasmuch as those reported here were more common during growing season.

why they used particular values of magnetic field rather than others when more than one was reported during a given year. The measurement values between about 1.2 and 2.6 mG all come from a single year (1991). The authors' conclusion that there is an effect on tree growth at about 2.0 mG is not warranted unless they define their field measurement more carefully, provide a clear rationale for it, and find consistent results from more than one growing season.

CONCLUSIONS REGARDING USE OF EXPOSURE DATA

Exposure data were inadequately or inappropriately used in a number of studies. The studies that used short-term response measures should have used, but did not use, transmitter on and off times to determine exposure, in addition to the IITRI division of sites into treatment and control. With a few exceptions (the upland-flora study and the wetlands study to a lesser degree), the ELF-EMF data provided by IITRI were used in site selection but not in any way to attempt to establish an exposure-response relationship. Those data should have been used more than they were. In the one case of a search for an exposure-response relationship, the data were used in a way that was not clearly related to actual exposures over the duration of the growing period. Thus, without further justification, any conclusions about low-level magnetic-field effects on tree growth are not warranted.

IITRI was successful in generating data sets on exposure to ELF EMFs for each study at the two antenna facilities. As far as the committee knows, the data sets were made available to researchers, and additional data were generated at particular study locations on request. It is not clear, however, whether IITRI followed up the generation of the ELF-EMF exposure data with assistance to each monitoring study in using the information. And when the researchers began to state their findings in annual reports or at the annual meeting, it is not apparent whether IITRI understood the extent of use of the exposure information by each study. The committee's evaluation of the separate studies indicates that use of the exposure information was often inadequate, inaccurate, or inappropriate. That should have become obvious to IITRI and external reviewers when they read or listened to reports from each study each year. The committee wonders why more guidance was not given to each study investigator in using the exposure information to ensure that the response results reported in the final reports of each study were based on appropriate use of the exposure data. IITRI should at least have required that an EMF-exposure expert work closely with each study until the study leader understood the types of data available, their variability (considering the vagaries in antenna operation), and how they might be applied to gauge responses of the selected ecological or biologic variables.

STUDY-SITE SELECTION

The original request for proposals (RFP) for research regarding the effects of ELF EMFs on biologic systems emphasized selection of study sites so that appropriate levels of ELF EMFs existed at control sites versus treatment sites. Considerable attention was also paid to the importance of matching sites so that any major differences uncovered would indeed reflect responses to ELF-EMF exposure, as opposed to uncontrolled effects due to contrasts in soil composition, chemistry, or vegetation. It would have been useful at the outset to have an integrated research plan in mind to guide site selection. The absence of such a plan meant that less than full value was obtained from the research as a result of segregated selection of possible ecosystem effects. In addition, the lack of common exposure levels at sites limited the possibility of integrating results. (See the section later in this chapter on lack of integration.)

THE PRACTICAL PROBLEM OF SITE SELECTION

Given the unavoidable heterogeneity of soils and vegetation typical of the Michigan Upper Peninsula, most researchers did as good a job as possible in matching sites. For instance, the earthworm and soil-arthropod sites were well matched with respect to soils and arthropods, but not with respect to earthworms. A better match with respect to earthworms could probably not have been attained. In the aquatic study, the researchers sought an upstream control site but could not find one, because none that was of the same stream order and physical attributes as the treatment site existed. In general, so many conditions had to be met for site selection that perfect matches were impossible.

ADJUSTING THE RESEARCH PLAN TO PROBLEMS WITH SITE SELECTION

Once the site-selection process was undertaken, it quickly became obvious that ideal matches were going to be difficult to find. Numerous researchers attested to that in their response to written questions from the committee. Two modifications of research design should have been considered upon encountering the practical obstacles to perfect site matching. First, for a small subset of critical variables, it would have been valuable to pursue spatially extensive comparisons of multiple sites (control versus treatment), thereby gaining the inferential power of multiple independent samples. The more-extensive design would require less effort at each site, although there would have to be a tradeoff in reduced number of variables measured. Second, for the intensive paired comparisons of sites, more critical thought might have been given to what processes, organisms, and observations would be most valuable, given some site mismatches. For instance, in the soil arthropods and earthworms study, the work on the dominant earthworm at the treatment site alone could never yield results that could clearly be attributed to ELF-EMF exposure, because there was no way of separating ELF-EMF effects from other factors without a control site at which the same earthworm was studied.

CONCLUSIONS REGARDING SITE SELECTION

Researchers faced difficult problems in selecting sites because of heterogeneity in the environment. Additional money and time for the sampling of more sites and some rethinking of experimental design and statistical inference would have helped to address some of these problems. For example, the program would have been improved by studying fewer variables at more and larger sites and by eliminating studies with poorly matched treatment and control sites. However, as with any environmental impact assessment, it would be unrealistic to expect the Navy's monitoring program to conform fully to an ideal experimental design.

PSEUDOREPLICATION

In many of the studies, the main effect of interest, namely the effect due to the presence of ELF EMFs generated by the antennas, was pseudoreplicated (that is, not truly replicated, as described by Hurlbert 1984) in that there was only one site for each level of exposure. Even when more than one site was available, treatments were not randomly interspersed so that background effects of soils, climate, etc., were equally (or at least randomly) distributed among all treatments and controls. Therefore, the experimental data provide an estimate of variance of responses studied within each site, but not the variance due to treatments across sites. The effects of the antenna on response variables are therefore confounded with the background effects of the different soils, climate, etc., on each site, and the two cannot easily be separated. This problem arises in the litter decomposition and microflora studies, the aquatic ecosystem studies, the upland-flora studies, and others. The wetland studies and the bird community studies avoided this problem by having replicate sites within different treatments and replicate plots or transects within each site. In the latter studies, the variance of the response variable of interest could be separated into two components: variance within each site and variance due to treatment or its absence. In the pseudoreplicated studies, that was not possible, because there was only one site per treatment, so the variance recorded can, strictly speaking, be attributed to only within-site, not between-treatment, effects.

Sometimes pseudoreplication is necessary for logistical reasons. When pseudoreplication is unavoidable, the generalization of treatment effects to other sites might be justified (with caution) if it can be demonstrated that the sites chosen for each single application of the treatment are not substantially different from each other at the outset and are at or very near the modal values of other environmental factors thought to affect the response variable of interest. That type of pretreatment survey was performed only for the upland-flora studies. In contrast, the litter decomposition studies are seriously compromised because there were large differences in decay rates between sites before the antennas were in operation. The antecedent site effects obscured the potential detection of treatment effects.

The acceptance of all further conclusions must proceed with those caveats in mind. The caveats are generally not clearly stated anywhere in the monitoring-program reports. The danger with pseudoreplication is in committing a type II error—accepting the null hypothesis (no effect) when it is, in fact, false. Such an error could arise, for example, if the variability within sites or between matched sites is within the range of responses imposed by the treatment. That should not be taken to imply that the effect of the antenna is small, as implied in some reports. Rather, it is not possible to separate the effect of the antenna from effects of other environmental factors without replication both within sites and across sites. In many cases, it is impossible to calculate the probability of a type II error because it depends on an independent estimate of differences between treatment and control, which requires replication of sites, not simply of plots within sites. The putative treatment effect is confounded with the site effect, so the differences between treatment and control cannot be attributed solely to the antenna, inasmuch as they are not independent of pre-existing effects and confounding site effects that continued during the experiment.

The extensive use of ANOVA and other linear models (including regression techniques) in the ELF study requires some strong assumptions about the distribution of the data, namely, that effects are linear in the scales chosen, that variances are constant, that error terms are independent, and that residuals are normally distributed. Those assumptions were not usually tested by the ELF researchers, with few exceptions such as the study of the effects of the antenna on bird populations. Furthermore, when the assumptions were considered, the investigators seem to have misunderstood them and to have applied other statistical analyses that might not have been the most appropriate. An example of the latter problem is the rejection of exponential-decay models in the litter-decomposition experiments in favor of covariance models that are more difficult to interpret. A further, troublesome aspect of the analyses is the nearly complete absence of any quantitative discussion of the effects of statistical bias, which could well dominate the role of random variation. The lack of quantification of statistical bias is exacerbated by the pseudoreplication or even lack of replication in many of the experiments.

SPECIES SELECTION

The Navy's original plan for an ecological monitoring program recommended that species (or related species) be studied that are reported to be sensitive to EMFs; are important ecologically, aesthetically, or economically; and can be reasonably monitored. These recommendations were largely met in these studies, as discussed below.

STUDY SPECIES

The diversity of species studied was considerable. Studies included species in most major taxonomic groups, including vascular plants (trees and shrubs), algae, slime molds, amebas, fungi, small mammals, birds, arthropods, and various decomposers. However, studies on small-mammal populations and on development of vertebrates (birds and mammals) had design problems, so information on this group is unreliable. Major groups of organisms that were not included were some nonvascular plants (e.g., moss), reptiles, and amphibians. In the wetlands study, a moss population was found to increase significantly at the treatment site, but this apparent response was not pursued, because moss was not a target species; this is unfortunate because the finding might be an indicator that moss is especially sensitive to ELF-EMF effects.

The species studied included types of organisms that had been reported to be sensitive to EMFs in previous laboratory or field studies, including slime molds, vascular plants, earthworms, birds, and bees. This coverage was very good. Little information exists on the EMF sensitivity of most of the particular species in the site, so species related to those exhibiting an effect were usually studied. For example, native bees, rather than honeybees, were studied because honeybees cannot survive the winter in this area.

Other species were usually well justified on the basis of their potential ecological importance. No study used economic or aesthetic importance. The most common criterion of ecological importance was abundance, which is reasonable because abundant species often exert a large effect on ecosystem processes and on other species that depend on them for food and performing valuable ecological functions. For example, the most common tree and shrub species were studied. The ecological importance of other species was based on their potential functional importance to the ecosystem. For example, periphyton species are valuable as food for many other species and as bioassays of water quality. Decomposers are crucial to nutrient cycling. Streptomycete populations are associated with decomposition and nutrient cycling.

Most study species could be adequately monitored because they were abundant, although some studies were limited by small sample sizes. No studies focused on rare species, although some might inadvertently have been included in aggregate variables such as bird censuses. In the wetlands study, rare sedge species were dropped when it was recognized that sampling might harm them.

The omission of rare species is problematic. Some rare species, such as predators and keystone species, can exert major effects on communities. Potentially endangered species were not studied. It is not known whether rare populations at the edge of their range are more or less sensitive to bioassays of additional stresses than abundant species. It is important to note that studying rare species would have been useful as studying more-common species in an attempt to find any type of effect.

The ability to measure response variables determined the choice of some species. For example, in the wetlands study, several species were dropped when it became clear that stomatal resistance could not be easily monitored.

In a few studies, species were chosen to provide interesting comparisons and generalizations. For example, in the upland litter-decomposition studies, the leaves of fast- and slow-growing tree species were compared. In the wetlands study, trees, shrubs, and herbs were compared. However, the limited number of species from each group precluded generalizations about life forms. More such comparisons would have been valuable in these studies.

CONCLUSIONS REGARDING SPECIES SELECTION

Overall, species selection was commendable. The species studied included a broad range of organisms with potentially different responses. That is important for detecting potential ELF-EMF effects on ecological systems that contain a wide diversity of organisms. Representatives of most taxonomic groups that have been reported to respond to EMFs were included. Most other species were well justified on the basis of their abundance and potential ecological importance. Aspects of concern include the lack of studies on reptiles, amphibians, and nonvascular plants; the lack of focus on any rare species; the decision not to pursue investigations regarding the population increase in the moss species; and the lack of reliable studies on the development of birds and mammals.

RESPONSE-VARIABLE SELECTION

The first priority of the original RFP was a study of bird migration and nesting success because birds have been reported to use magnetite for orientation. However, all the studies addressing that priority had unreliable or weak tests of the response variables. Bird migration was not examined, because critical outside reviews caused the proposed study to be canceled. Instead, local homing and navigation by resident birds were studied. The negative results are questionable because data on antenna operation during the specific periods of study were not used in the analyses. A similar problem is raised by the study of nestling development in tree swallows. Actual field strengths during critical developmental periods were not considered, so the negative results might reflect the lack of exposure rather than the lack of an ELF-EMF effect. The study of bird populations also relied on a weak test: the treatment site was far from the antenna and had very low exposure levels.

The second priority of the original RFP included studies of soil microbiology and ecology, plant ecology, and insect populations and behavior. These were all examined. Soil microbiology was examined as litter decomposition in three studies (litter decomposition and microflora, wetland studies, and earthworms and soil arthropods) and as streptomycete populations, which are good indicators of microbial community activity (upland flora). Plant ecology was examined in the studies on upland trees and wetlands. Insects were examined in the studies on soil arthropods, aquatic insects, and pollinating bees.

The third priority included water quality, fish ecology, reproduction, fertility, and biorhythms. In the aquatic project, water quality was monitored through periphyton responses; and fish abundance, size, growth, and movement were measured. Reproduction was studied in bees, earthworms, birds, and small mammals. No studies addressed biorhythms or fertility. No studies examined fluctuating asymmetries in development, which have been shown recently to be sensitive indicators of environmental stress.

The omission of mammalian responses in the list of original priorities is of concern. Although the actual studies included mice, the study was flawed, and results are inconclusive.

CRITERIA FOR RESPONSE SELECTION

The original RFP stated two major criteria for response selection: the ability to measure responses accurately (consistently) and the ability to detect differential responses amid variation caused by other factors.

The first criterion was usually adhered to for practical reasons. If accuracy could not be achieved, species or response variables were generally dropped. For example, several wetland species were dropped when stomatal conductance proved difficult to measure. Nitrogen fixation was also dropped in the wetlands study. The second criterion was often not met. In addition, many studies suffered from poor experimental or statistical execution and a lack of power that precluded finding significant effects.

To address the second criterion of detection, three major approaches that varied in effectiveness were used. First, some researchers chose response variables that they argued were unlikely to be affected by factors other than ELF EMFs. For example, leaf foraging by bees, rather than flower foraging, was studied because flowers varied among sites. However, other factors that could affect leaf foraging were not examined. In the wetlands-plant study, a potential membrane effect on stomatal conductance proved to be affected so much by local conditions (light and temperature) that a species unresponsive to sun was chosen; whether this species might also be unresponsive to ELF-EMF effects is open to question. In the slime-mold study, field ELF-EMF conditions were mimicked in the laboratory to the greatest extent possible to avoid confounding factors.

The second approach for increasing detection was to examine changes before and after antenna operation. That was effectively done in the soil-arthropod study by using BACI analysis and in the bird population study by using repeated-measures analysis. In some other studies, the response variable (mass loss in wetland plant litter, overwintering mortality in bees, and bird populations) was known to vary before antenna operation because of other factors. In those studies, researchers tested for a significant interaction between treatment and year; however, the weak correlation between year and ELF-EMF exposure intensity and the lack of power (replication) usually precluded finding any significant effects.

The third approach for increasing detection power was to incorporate other variables into predictive models. This is an effective method that was used in the upland-flora study. Individual tree growth was predicted on the basis of measurements of site, climate, and competitors. Residuals (discrepancies of the observed data from the predicted values) were then tested for significant effects of ELF EMFs. However, the choice of exposure measurements makes the conclusion of positive effects questionable.

Detection of possible ELF-EMF effects on short-term biologic responses depends crucially on the timing of the antenna operation relative to the experiments. However, this timing was apparently ignored in the majority of the studies although data on antenna operation were available. Because antenna power was variable (including zero during shutdowns), knowledge of the timing of exposure is critical to interpreting the results. For example, the orientation of birds flying over ELF EMFs was analyzed without regard to the actual operation of the antenna during each experiment. Embryo development in birds had a 4-day window when exposures could be effective, but antenna operation times were not considered. In the wetlands study, the individual measurements of stomatal conductance and water potentials of plants were apparently not related to antenna operation times. If ELF-EMF effects are immediate and reversible, no responses would be seen during temporary shutdowns. Negative results could reflect absent or highly variable exposures rather than no effect of ELF-EMF exposure.

Longer-term responses, such as growth of individuals and populations, will reflect the accumulation of ELF-EMF effects over time. Therefore, the variable timing of antenna operation is of less concern than for short-term responses. However, spatial variation in ELF-EMF intensities could be important for sedentary species. The upland-flora project was the only study in which variation in individual exposure rates was measured and used in the analyses. In an extreme case, slime molds were removed from the ELF-EMF exposure, and effects on the next generation were studied. For more-mobile species, exposures would be impossible to determine, but their movements probably average out any spatial variation in exposure.

Population responses were examined for periphyton, soil microfauna, insects, birds, and small mammals. These species were good choices because most have relatively short generation times and populations could show responses over the duration of the studies. However, three systems that showed potentially strong population responses were not adequately studied. In the wetland study, moss increased but was not studied, because it was not a target species. Slime-mold populations appeared to be highly sensitive in initial studies, but population growth was not studied in the field site; instead, later generations taken into the laboratory were examined for responses. A statistically significant increase in chlorophyll-a of periphyton was compelling, but no followup laboratory studies were done, apparently because of researchers' resource limitations.

In general, population responses were statistically nonsignificant, but this does not necessarily mean that biologic effects were nonsignificant. For most of the population studies, statistical power of the tests was low, so the ability to detect population response was small. For example, in the bird-population studies, survival of nestlings could be reduced by 10-30% without detection of the reduction in population changes, because most adults probably live 2-6 years. Also, populations could be maintained by dispersal from control sites.

Effects on reproduction, growth, or mortality can be more sensitive measures of population growth rates than population densities. More effort to measure these variables accurately could have been informative. Many of the studies that were done were flawed. The study of nestling development was flawed by uncertainty about antenna power levels during critical periods of development. Similar studies on mice were too limited to yield strong conclusions. Growth rates of trees were studied, but the predictive model was flawed. One of the strongest results, increased overwintering mortality in bees exposed to ELF EMFs, was dismissed by the researchers, perhaps too readily. In the slime-mold study, the mitotic cycle had been reported to be a sensitive variable but was dropped from the study.

Responses of communities, measured by such indexes as diversity, were examined in the studies on soil arthropods, aquatic systems, and bird populations. Ecosystem responses were examined as nutrient uptake and decomposition in the plant studies. None showed significant effects. However, such aggregate variables can mask impacts on underlying processes, especially if there is any compensation; for example, one species replaces another, and increased reproduction balances increased mortality.

WEAKNESS IN THE GENERAL RESEARCH DESIGN FOR RESPONSE VARIABLES

One weakness of the research design for examining responses was the lack of emphasis on understanding possible mechanisms and using mechanistic models. Some researchers did justify their response variables on the basis of known possible mechanisms. For example, the claim that EMFs can affect membrane potentials was used to select stomatal conductance and foliar composition as responses to measure in wetland plants. More consideration of mechanisms could generate more-specific experimental tests or predictive models.

Another weakness of the general research design was the lack of establishment of relationships between exposure or dose and biologic responses: exposures and dosage were poorly understood or not used in the analyses; and if responses were nonlinear with respect to exposure or dose, some effects might not have been detected. Most studies did not address either problem. An exception was the upland-flora study, in which exposures were estimated for individual trees; results indicated a slight stimulation of tree growth at moderate levels of ELF-EMF exposures, but this effect might have been spurious and caused by misuse of exposure measurements.

CONCLUSIONS REGARDING RESPONSE-VARIABLE SELECTION

Many responses, including short-term and long-term responses, were measured, and that is commendable for detecting potential effects at different levels. However, the ability of the studies to detect possible ELF-EMF effects was generally weak. Many studies, especially those on birds and mammals, were flawed in ways that reduced the likelihood of detecting statistically significant responses. Timing of exposure was not related to measurements of short-term responses. Possible small effects would have been difficult to detect, given the lack of power and the occurrence of confounding variables. The term "small effects" is used in this report to refer to ecological effects whose magnitudes are not likely to exceed those expected from normal perturbations over the short term. The lack of information on functional response relationships to ELF-EMF exposures might have reduced the researchers' ability to decide which organisms and response variables are most likely to exhibit effects of the ELF antenna. The lack of reliable information on vertebrate responses is of special concern. The statistically significant population responses of periphyton, moss, and pollinating insects might be worth following up in more-controlled laboratory studies.

STATISTICAL POWER

The power of a statistical design reflects the likelihood that an experiment will be able to detect the presence or absence of a treatment effect. The importance of adequate power is simple: a study with low statistical power will not be able to accept or reject the null hypothesis with sufficient confidence. In the absence of adequate statistical power, the studies in the ecological monitoring program are uninformative at best. At worst, a potentially important effect might be missed and replaced with an unwarranted sense of confidence that the null hypothesis of no effect has been substantiated.

The committee noted lack of adequate statistical power as a problem that arose in more than one study. In the small-vertebrates study, an early decrease in sample size led to a decrease in power to 70%, and a large number of variables were eventually analyzed with statistical power of less than 30%. The power of statistical tests in the soil arthropods and earthworms study was low because of the sampling scheme (few pitfall traps), large differences between sites, and large variability in the data. It was impossible to estimate accurately the statistical power in several studies, including Michigan and Wisconsin birds and litter decomposition and microflora, because of pseudoreplication and unclear exposure relationships. Reanalysis might improve a number of these studies (see Chapter 5); others are unsalvageable because of flaws in their design or execution. Some studies, such as the pollinating-insects study, did a good job of calculating and discussing the minimal detectable differences and the power of statistical tests.

To quantify the likelihood of detection, or power, one needs to describe three aspects of a study: the experimental design and sample size, the rules by which an effect will be declared statistically significant (as distinct from biologically significant), and the magnitude of effect that will be assumed to result from the experimental intervention.

For example, consider the following hypothetical scenario chosen for its relevance to the ELF ecological monitoring program:

- The goal is to compare 20 pairs of nesting birds in a control plot with 20 pairs of birds at one of the treatment sites. The variable to be examined is the number of surviving hatchlings per nesting pair at some specified time.
- The variable (number of surviving hatchlings) will be assumed to follow a Poisson distribution. The statistical rules are (1) the null hypothesis of no difference in the means of the numbers of surviving hatchlings and (2) rejection of the null hypothesis according to a two-tailed test at the 95% significance level.
- The statistical goal is to be able to detect a 20% or greater change in mean number of survivors, under the assumption (justified from prior data) that the mean control number is 3 surviving hatchlings per pair. (Typically, the magnitude of the change to be detected is dictated by considerations of biologic significance.)

For each case of exposition, we will also assume that the hatchlings are independent of each other. Although this assumption is not correct, it eases the presentation without compromising the illustration with respect to statistical power. That scenario provides sufficient information to compute the statistical power of such an experiment. Statistical power corresponds to the probability that a significant effect will be observed if the mean number of survivors at the treatment group differs from that at the control site by 20% or more, provided that all other assumptions have been satisfied.

It is vital to consider confidence intervals, that is, to consider the results of such experiments not as providing only two mean values with an associated p value for their difference, but as providing the differences to be expected if the experiment were repeated many times. For example, assume that the results of the above experiment are that the mean number of hatchling survivors in the control group differs from the mean number in the treatment group by 22%, with 95% confidence limits of 8% and 150%. If the experiment were repeated many times, one would expect the mean difference between treatment and control groups to be in the range of 8-150% in 95% of the repetitions. Because the confidence interval does not include zero, the null hypothesis is rejected at the 0.05 level of significance. The width of the confidence interval implies that the effect could be uncomfortably close to the null (8%) or could be quite large (150%); that is, the experimental design and sample sizes have led to very imprecise estimates.

What does the outcome of the experiment tell us if the power of the design is 0.9? It indicates that if there is indeed an effect of 20% or greater, 90% of the time the confidence interval will not include the null hypothesis. It provides us with some reasonable bounds on the uncertainty. An alternative experimental design with a power of 0.3 or less (like those reported in a number of experiments in the ELF ecological monitoring program) tells us that 70% or more of the time the confidence interval could include the null hypothesis. Such an experiment offers no bounds on the uncertainty, and it can reasonably be questioned whether such an experiment should have been performed at all. In this context, it is useful to note that it is common to require that results be expressed in terms of confidence intervals.

RESPONSE TO REVIEWS AND CRITIQUES

Research teams associated with the ecological monitoring program received annual comments from reviewers beginning in 1982. In its evaluation of individual studies, the committee discovered that there was a great deal of variation in how the researchers responded to the reviews of their annual reports and presentations. Some review comments appeared to have been taken seriously and to have led to modification of research designs or report presentations; other comments appear to have been taken lightly or ignored. The committee did not attempt to determine, in a systematic manner, the extent to which the researchers considered the review comments to be appropriate. However, the committee found several instances in which the peer-reviewer comments raised valid and important questions. Some of these issues have been addressed in this report.

In a number of cases, reviewer comments were responded to in a satisfactory manner. The initial study on Michigan and Wisconsin birds was canceled in response to negative reviewer comments, and a new proposal was accepted. The wetlands researchers modified their study design to accommodate reviewer comments and provided explanations if changes were not made. Upland-flora researchers took peer review seriously and addressed the concerns of their reviewers.

Several research teams responded only partially to reviewer comments. The research team for the litter-decomposition study made minor changes in experimental technique that were suggested by reviewers. However, more difficult problems involving theoretical issues were only partially considered, if at all. The researchers appeared not to understand some issues fully. In the pollinating-insects study, most reviewer suggestions were heeded and contributed greatly to the quality of the research. However, suggestions on using a BACI analysis with covariates and on increasing replication were not followed. Small-vertebrates researchers were generally responsive to peer reviewers and made several improvements in the study and in the clarity of the final report. However, reviewer concerns regarding low statistical power and lack of data archiving were not addressed.

Every year, IITRI received annual reports on all the studies and held a meeting at which each study was presented or discussed. IITRI organized outside reviewers to review the annual reports and to comment on the progress of each study. Some of the external reviewers' comments were very critical. Such criticism should have been a clarion call to IITRI that something was seriously wrong with the design or progress of the criticized study. It is not apparent to the committee that IITRI followed up on reviewer critiques of studies; followup seems to have been left to individual investigators. Consequently, research design, analytic techniques, or interpretation that needed improvement or correction according to external reviewers were often left unattended. That lack of attention to reviewers' comments should have concerned IITRI management. IITRI should have established a regular internal review process to ensure that each study adequately addressed external criticism, even to the point of having external reviewers comment on responses to their criticism. There are also instances (e.g., data archiving) where critiques and suggestions of the peer reviewers were addressed to IITRI managers directly. These critiques appeared to have been ignored by IITRI.

APPROPRIATENESS OF INTERPRETATION

Scientists are well acquainted with the potential for bias in conclusions based on a given set of data. Preference for a particular outcome could result in an interpretation favoring that outcome, and this could occur even in cases where the persons making the judgment believe that they have done so without bias. (Such a tendency toward bias could also be present in the subjects of experiments involving humans. Those considerations have given rise to so-called double blind experimental protocols, in which neither the subject nor the experimenter has knowledge that will allow such potential bias.)

In several studies of the Navy's ecological monitoring program, modest but significant differences were observed between data collected at treatment sites and data from control sites. Researchers conducting the studies concluded that five of these potential effects were due to factors other than the ELF antenna. Without attempting to judge whether any of those interpretations suggested a predisposition to a particular outcome, it is important to consider whether the conclusions were established with a credible scientific basis. In the course of the committee's review and discussion of the researchers' final reports, concerns arose about the scientific credibility of some of the conclusions.

Differences between treatment sites and control sites that were dismissed by researchers and by IITRI as not being clearly related to ELF exposure were the increase in bee overwintering mortality, the reduction in leaves per bee nest cell, accelerated litter decomposition, earlier eye-opening in mice, and depressed earthworm reproductive rates. The committee believes that some of these observed differences were dismissed too readily as alleged artifacts of environmental variations or experiment design. In the pollinating insects study, the final conclusion that ELF-EMF effects are absent or minimal is not fully justified by the data presented. Similar concerns were expressed by earlier reviewers, who concluded that ELF-EMF effects were demonstrated. In the litter-decomposition study, the provisional conclusion of negligible effects of the antenna appears to be based on a combination of small differences in mean values and the large variation in the data, not on the validity of the results. Even a 5-10% change in leaf-litter decomposition rates can have a major impact on soil characteristics (McClaugherty et al. 1985). Although dismissal of the five possible small effects described above might be correct, the committee suggests in Chapter 5 that the results of some studies be reanalyzed.

Another concern has to do with the issue of multiple comparisons and the observation of statistically significant results due to chance alone. For example, when researchers perform 100 tests at the 5% level of significance, one expects to find, on the average, five positive values (that is, to reject the null hypothesis) because of chance alone. In the very large number of statistical tests performed in the monitoring program, it would be surprising if no statistically significant findings were reported. Therefore, the issue is not whether significant results emerged from time to time, but whether the number of such events was larger then expected. That issue was not examined systematically as part of the ecological monitoring program.

DIFFERENT METHODS FOR SIMILAR ORGANISMS

The broad range of studies in the program often resulted in examining the possible effects of the antennas on similar organisms or processes, but with somewhat different protocols. For example, litterbag studies of decomposition were performed in uplands, wetlands, and streams, but these studies were performed by different teams, often from different universities or institutions, and were initiated at different times. Under those circumstances, it is understandable that different methods might be used. The upland litterbag studies used leaf litter from dominant tree species, and bags were sampled at about monthly intervals. In the wetland studies, decomposition experiments began with cellulose strips and switched to leaf litter from *Ledum groenlandicum*, a common shrub. However, the samples in the wetland decomposition studies were taken only annually. Annual sampling precluded fitting different decay models to the wetland decomposition data, as was possible in the upland studies. It is not possible, therefore, to analyze all the decomposition data with a common statistical or mathematical model.

The use of different methods does not necessarily negate conclusions from any one study. Indeed, there might be sound scientific or logistical reasons for using different methods in different situations. For example, achieving greater precision to meet the objectives of one study might entail different methods from those in another study. But the use of different methods prevented researchers from making valid cross-site comparisons and thereby impeded the realization of the full potential of integration across sites and organisms. For example, in the upland-forest decomposition studies, litter from various species was placed at each site as an "index" material. In the wetland-decomposition studies, cotton strips were used as index materials. Litter in the upland studies was sampled monthly for mass-loss estimates, but in the wetland studies it was sampled annually. There does not appear to have been much discussion of coordinating methods before the experiments began. Therefore, the inability to integrate properly across sites and organisms because of the use of different methods is an unfortunate after-effect of the execution of the work, rather than a tradeoff that was intentionally made for the sake of greater precision within any one study. The collection of studies as a whole would have benefitted from the specification of a small number of hypotheses at the earlier stages of study design.

LACK OF INTEGRATION AMONG STUDIES AND SYNTHESIS OF INFORMATION

As noted in Chapter 1, the original RFP for studies of the effects of ELF EMFs on biologic systems in the region of the ELF communications system antennas was developed by IITRI on the basis of a monitoring program outline from the Navy, previous research, information from state agencies and the U.S. Forest Service, and comments on the Navy's draft environmental-impact statement. Recommendations from the National Research Council report (NRC 1977) on potential effects of ELF EMFs also influenced the RFP. The RFP requested proposals addressing the responsiveness of select groups of organisms to the environment created by the EMFs produced by the ELF antennas. The list included such groups as mammals, birds, invertebrates, plants, slime molds, and amebas. There was only a slight suggestion in the RFP that studies of effects were to be in an ecosystem context; rather, the emphasis seemed to be on a population approach.

SELECTION OF PROJECTS

In the process of selecting the ecological studies, there does not appear to have been an attempt to fund research teams that were sufficiently close together to encourage interaction. That could have been the result of selection of the best proposals from researchers from widely scattered institutions, disjunct expertise within each of the selected organismal groups to be studied, or a response to "requests" to diversify the selected institutions regionally to satisfy "political distribution" of research funds. Regardless, the competitive process and the selection of studies from many institutions created a situation that did not encourage or permit much integration of study locations or data.

SELECTION OF STUDY SITES

Investigators of each of the ecological monitoring studies were allowed to establish their own study sites as long as they satisfied IITRI's exposure criteria. There do not appear to have been any attempts made by IITRI to encourage proximity or overlap of study sites. The existence of antennas in Wisconsin and Michigan precluded designing research projects that would always be associated with each other, but projects at one or the other of the antenna could have been closely aligned. Different starting times and basic hypotheses for projects and selection of investigators from many different institutions might also have created a condition not conducive to integration efforts. The lack of integrating study locations also resulted in a wide range of exposure levels for both experimental and monitoring studies in this project. There apparently was one exception to the isolation of study sites and efforts: the use of the forest site at the Michigan antenna for the decomposition study.

It could be argued that the diversity of studies was so great as to preclude any integration. For example, forest, wetland, and aquatic sites and study sites for invertebrates, birds, amebas, and slime molds might have been considered sufficiently different or spatially separated to prevent integration of studies. However, within a few kilometers around each antenna, it seems that it would have been possible to locate all the ecosystems studied and within these to establish study locations for specific organismal groups. By doing that, the planners of this project would have taken a landscape approach and encouraged an understanding of the interactions among the various systems and the organisms associated with them.

Integration of the studies would have permitted project planners to use workshops and other forms of communication to generate a synthesis document, which could have demonstrated how, for example, forests alter ELF-EMF intensities and thereby modify exposure of associated organisms, especially subordinate species, such as those which influence decomposition rates. Other examples of possible synthesis might have related wetland systems adjacent to streams with the responses of aquatic attributes or compared responses of invertebrates within different but integrated study areas. Many other examples could be given.

CONCLUSIONS REGARDING INTEGRATION OF STUDIES AND SYNTHESIS OF INFORMATION

In the processes used for selection of studies and study sites, there appears to have been little recognition of the possibility that a response of one attribute of an ecosystem (such as an organism or process) could influence other ecosystem components and that information about related responses might reinforce or undermine conclusions about responses taken one at a time. Instead, possible responses were considered as isolated events, that is, outside an ecosystem or integrated context. Such a perspective might have caused the managers of the ecological monitoring program to establish study site selection requirements based only on exposure levels and research impacts on limited populations of sensitive species. The committee notes that it would be infeasible to synthesize the data on response variables because they represent diverse aspects of ecosystems and because in most cases the measured responses were insignificant. In addition, looking for these interactions and developing a synthesis document might take too long and thus be outside the purview of the ecological monitoring program. The committee also notes that there might be some value in use of varied approaches, given that so little is known about the effects of ELF EMFs on ecosystem components and processes. Nevertheless, recognition of interactions among ecosystem components and encouragement of integration among studies with full development and application of appropriate statistical approaches should have been guiding principles in the early research design. That would have given the research community the opportunity to synthesize the extensive information generated over many years of study, even outside the funding of IITRI. Both the advisory committee and the monitoring program's management team were remiss in not including recommendations, and perhaps requirements, for integration and synthesis in the RFP.

An additional problem with the lack of integration and synthesis is that there was virtually no opportunity to follow up on the most-positive research findings in lieu of continuing with less-promising research. For instance, the finding of a statistically significant increase in chlorophyll-a in response to ELF EMFs in streams was never followed up with the obvious laboratory studies, because of lack of funds, even though the researchers recognized it as a potentially important finding that needed laboratory confirmation. Conversely, earthworm comparisons between sites that differed vastly in their initial earthworm fauna were pursued, even though findings were doomed to ambiguity (because of poor comparability of sites with respect to the earthworms). The funding and contract basis of these studies locked the program into an inflexible pattern of support for research without matching the funding to a continuous review of the results. If there had been consistent integration and comparison of findings for the different studies, the overall research effort could have been improved for the same amount of money. With better integration, there could have been more pursuit of promising results, the hallmark of good research.

By failing to integrate the studies of different species and ecosystem processes, this large-scale effort largely surrendered the possibility of detecting small changes in interactions of components and gave up the major advantage of such large-scale research. Given the absence of synthetic overview, this might just as well have been many isolated studies of one variable at a time. The project thus missed an excellent opportunity to conduct pioneering ecosystem-level research that spanned physiologic, population, and ecosystem responses. All ingredients for such research were in place—but no one put the ingredients together.

DATA ARCHIVING

The final and annual reports do not contain information on archiving of data generated by the Navy's ecological monitoring program. To understand the current state of data archiving, the committee spoke with several researchers directly. The results of these telephone conversations indicate that each group of researchers used its own protocol for archiving data and that the quality, durability, and accessibility of these protocols differed dramatically. For the wetlands study, data do not exist in electronic form but are available in the original notebooks and in the annual and final reports published (F. Stearns, formerly of University of Wisconsin-Milwaukee, personal commun., 1996). For the soil-ameba study, biologic data are recorded in notebooks, and environmental data are available on computer printouts. Although many of these data were analyzed in electronic format, the software used was old and the data are probably not retrievable at this point (R.N. Band, Michigan State University, personal commun., 1996). All data from the pollinating-insects study are contained in a relational database, RBase. However, the data would not be useable with the limited documentation available, and anyone who wanted to use them would need to speak to the researchers to have the variables and peculiarities of the database explained (K. Strickler, University of Idaho, personal commun., 1996). Data from the study of bird populations in Michigan were archived with a relational database, Paradox. With some documentation, others would be able to use the files. These data were archived in a standardized format used at the researchers' institute (J. Hanowski, University of Minnesota, personal commun., 1996).

The small-vertebrate study is worth noting in this context because the large, ambitious study yielded voluminous data. It is important to separate the value of the study as a purely scientific investigation from its value as a targeted effort to ascertain whether the Navy's ELF communications system had deleterious effects on the neighboring ecosystems. As a purely scientific investigation, it has yielded mostly high-quality data with incomplete statistical analyses. Some of the flaws identified in the evaluation of the study in Chapter 3 could, in theory, be remedied by analyses that used a framework both physiologically appropriate and statistically sound. Such analyses are not likely, in part because of decisions on the part of the IITRI management team regarding data archiving. The management team failed to observe elementary practices of data management that would have yielded a documented archive of data suitable for re-examination. The latter is an egregious failure, inasmuch as a peer reviewer of the entire monitoring program, made the following pointed comments before the 1988 contract-renewal process (letter dated May 5, 1987):

> All participants should be sending YOU hard copies and floppy disks (IBM compatible) right now. All data sets should have excellent documentation and you should have copies of it. *I was a little disturbed that some participants take a lackadaisical view of their data sets. . . . Each investigator should consider data management as an obligation. For IITRI it is an essential."* [Italics added]

Apparently, the recommendation was not followed, and documented archiving of data was not undertaken.

The Navy ELF ecological monitoring program supported 11 long-term studies. Those studies generated extensive data sets on biologic response variables and environmental conditions. Most of the studies continued for 5 years or more, and each produced annual reports of progress and information gathered. From a review of the overall program, it appears as though all that was expected from each study was an annual report, presentation or attendance at an annual meeting of monitoring-program investigators, and a final report that discussed responses of selected variables and drew conclusions. To meet those expectations, the investigators in each study must have developed extensive sets of information and organized them in a fashion that allowed analysis. Obviously, the data sets include working field notes, tables in electronic and possibly paper form, and results of analyses. As far as the committee is aware, those data sets were kept by individual researchers and not transferred to IITRI. No protocol seems to have been established for the request for proposal (RFP) or later for formatting, documenting, or reporting the data. After millions of dollars had been spent on a monitoring program that could be used for further understanding of the monitored ecosystems, the resulting information appears not to be readily available; if available, it is not in a uniform, user friendly format; and there is no common location to which an outsider can address requests for information. Responsibility for the lack of archiving and of planning for long-term availability of the monitoring information appears to rest with IITRI's management of the program. All aspects of archiving of monitoring data should have been designed as part of the RFP, and researchers in all the monitoring studies should have been required to submit their data sets, in the appropriate format, at regular intervals or at least at the completion of each study.

5

Overall Conclusions and Recommendations

THE NAVY'S ECOLOGICAL MONITORING PROGRAM reported no obvious adverse ecological effects, such as unusual changes in species populations or large-scale mortality of trees or other organisms, as a result of operation of the ELF communications system during the period of the monitoring program. The monitoring program also did not detect any small effects with well-defined consequences, such as decreases in reproductive fitness, that would be likely to result in major effects in the future.

The committee agrees with the general findings of the Navy' ecological monitoring program, within the limitations described in this report, that the researchers' observations provide no evidence of statistically significant, widespread, adverse effects of EMFs associated with the ELF antennas. For example, there is no evidence of short-term impacts of EMFs associated with the ELF antennas on bird populations, although effects within roughly 50-100 m of the antennas would not have been detected. The effect of antenna operation, if any, on leaf-litter decomposition processes and the microbial community is most likely much smaller than existing natural spatial and temporal variation in the forests. The total operation of the antennas seemed to have no major detrimental effect on upland flora beyond background environmental fluctuations. No experimental data indicated a statistically significant effect of ELF EMFs on the movement of dragon fly larvae, the colonization of leaf litter, or the movement of fish. However, the power of the experimental design and statistics probably could not have detected small effects. Effects of ELF EMFs on soil arthropod populations were small compared with the total variation of the processes measured over the 11-year study. No major effects were detected in soil ameba populations. For several studies, either the results were ambiguous and require further evaluation or they were uninformative because of design or analysis problems (see discussion below).

ECOLOGICAL EFFECTS

The term "small effects" is used in this report to refer to effects whose magnitudes are not likely to exceed those expected from normal perturbations over the short term. A drought is one example of such normal perturbations. The committee recognizes that small effects on populations, mediated through modest changes in response variables, might slowly compound and only later become apparent. Numerous flaws in the ecological studies—as designed, implemented, and interpreted—would have compromised detection of many such possible small effects of the antenna operation. As shown in Table 5-1, the individual ecological studies can be sorted into three categories: studies that the committee judged to be acceptable with qualifications (as discussed in Chapters 3 and 4), studies that might be acceptable after more information is obtained or data are reanalyzed, and studies that are unsalvageable because of serious flaws.

To some extent, the difficulties are inherent in field and ecological studies in which environmental fluctuations and site-specific differences, even with matched sites, cause changes that are large in relation to potential treatment effects and might obscure them. However, as discussed in Chapters 3 and 4, a number of studies were inadequate for determining small effects because of inappropriate exposure assessment, pseudoreplication, low statistical power, inappropriate interpretations of data, or problems with experiment design. As illustrated by the following specific issues and discussed in the remainder of this chapter, the results of several studies should be re-evaluated so that firmer conclusions can be drawn.

The environmental differences between treatment and control sites contributed to observed differences and might well have explained more of the variance in biologic responses than the ELF EMFs did. The inconsistency in litter-decomposition data from year to year illustrates the difficulty. The high

TABLE 5-1 Acceptability of Results from Response Variables of Individual Studies of the Ecological Monitoring Program[a]

Studies	Acceptable with qualifications	Might be acceptable with more information or analysis	Unsalvageable
Wetlands:			
Foliar nutrients	X		
Decomposition (moss growth[b])	X		
Stomatal resistance	X		
Slime molds		X	
Wisconsin and Michigan birds[c]:			
Bird populations	X		
Small vertebrates:			
Embryo		X	
Fecundity		X	
Homing		X	
Aerobic metabolism			X
Litter decomposition and microflora:			
Decomposition		X	
Microflora	X		
Upland flora		X	
Aquatic ecosystems:			
Algae (chlorophyll-a[b])	X		
Insects	X		
Fish			X
Pollinating insects (increased mortality[b])	X		
Soil arthropods and earthworms:			
Soil arthropods	X		
Earthworms			X
Soil amebas	X		

[a]Interpretation of study results should be qualified according to discussions in Chapters 3 and 4.

[b]Controlled laboratory experiments are suggested because of evidence of possible effects.

[c]Wisconsin bird study and Michigan bird study were combined by the committee for its evaluation.

variability between years in decomposition processes and in the microbial community studies within a site casts doubts on the possible small effects detected.

- The higher amounts of chlorophyll-*a* in algae in a stream at the treatment site were not accompanied by effects observed at higher trophic levels in the stream, as would be expected if there actually were an effect. Controlled laboratory studies might determine whether this effect is real or an artifact of a complex system.
- ELF-EMF exposure data were analyzed inappropriately in the tree-growth studies. An increase in tree growth due to ELF-EMF exposure was reported and publicized, but the committee's analysis of the primary data indicates misuse of the data. When the analysis is corrected, the growth effect disappears.
- The distinction between treatment and control sites might have been eliminated for experiments that had short-term response variables because the records of antenna on and off times were not consulted by the investigators. If the antenna was off, for example, during the incubation and embryo period of the developmental substudy of the small-vertebrate study, a treatment site would have become a second control site.
- Lack of replication is a problem in many studies and makes conclusions site-specific rather than broadly applicable. For example, if the conclusion of a study is that the antenna operation had no effect on the treatment site, it cannot be assumed that conclusion can be applied to other sites near the antenna.
- Most studies would have benefited from an understanding of mechanism or the use of mechanistic simulation models which might have improved statistical power, data interpretation, and prediction. The litter decomposition study would have benefitted from the use of a simulation model, although, as discussed in Chapter 3, the results would probably have been the same.
- A difficulty that runs through the investigations is the lack of statistical power in interpreting results and the misuse of statistical methods in eliminating some possible outcomes. The decrease in power from 90% to 70% and later 30% for some of the small vertebrate studies substantially weakened the researchers' ability to detect possible effects.
- The mortality of bees overwintering at a treatment site was found to be greater than that of bees at the control site, but the investigators argued without sufficient evidence that the observed effects are unlikely to have been caused by the ELF EMFs. Potential substantiation of a small effect was hin-

dered by provision of an alternative explanation. The data require reanalysis or reconfirmation.

- Slime mold cultures exposed to ELF EMFs in the field and moved to a laboratory for testing could well have recovered from any small exposure-induced effects in the long period between exposure and examination.
- The selection of a control site downstream of the treatment site for aquatic-ecosystem studies potentially contaminates the control site with treatment-site ELF-EMF induced effects and casts doubts on the results of the study.

IITRI'S ENGINEERING SUPPORT AND PROGRAM MANAGEMENT

CHARACTERIZING ELECTRIC AND MAGNETIC FIELDS

IITRI did a good job on the engineering aspects of the ecological monitoring program in characterizing the spatial and temporal characteristics of the electric and magnetic fields. The instrumentation for ELF-EMF measurements appeared to be well designed, well calibrated, and properly used. IITRI provided ELF-EMF exposure information to the researchers for each study. In addition, IITRI was responsive to requests from researchers for additional engineering support.

PROGRAM MANAGEMENT

When IITRI submitted its proposal to the Navy to develop an ecological monitoring program for the ELF antenna sites, it emphasized that its role would be not only management of budgetary components of an agreement, but also oversight of individual studies and ensuring of the quality and credibility of the overall program. This commitment appeared to give assurance that the monitoring program, which would generate extensive data from a wide variety of projects, would be closely supervised and that researchers in the future would be able to look at the results to understand how ecosystems in the upper lake states might respond to external perturbations.

In its review of individual study reports, as well as of the overall program of monitoring for possible effects of the ELF antennas, the committee discovered weaknesses in some aspects of IITRI management of the program.

CONCLUSIONS AND RECOMMENDATIONS 143

Chapter 4 of this report discusses weaknesses that were found to be common to many projects. Three of these weaknesses appear to have been caused by lack of adequate oversight. First, IITRI should have detected problems with use of exposure data through annual researcher reports. In response, IITRI should have provided more guidance for use of the exposure information and required that an EMF-exposure expert work closely with each study until the study leader understood the types of data available, data variability, and the best methods for applying the data. Also, there should have been greater involvement of an expert in broad-based applied statistics at the earliest phases in the design of this program's studies. Second, responsibility for the lack of archiving and of planning for long-term availability of monitoring-program information appears to rest with IITRI's management of the program. Third, IITRI should have established a regular internal review process to ensure that each study adequately addressed external criticism. These problems appear to have originated in poor early planning by IITRI or inadequate followup by IITRI as problems arose during the program.

RECOMMENDATION

The complexity of assessing the possible ecological impacts of ELF EMFs—especially given the diversity of the ecosystems and the variability of their locations and their distances from the antennas, (and therefore the variability of exposures)—made it extremely difficult to design and complete appropriate and comparable ecological monitoring studies. The following recommendation and next steps suggested in the subsequent section reflect the committee's understanding of such difficulties but also indicate the committee's concern for bringing this ecological monitoring program to an appropriate and fruitful conclusion.

Do Not Repeat the Monitoring Studies

Despite the weakness of the monitoring studies, the committee does not recommend that the field studies be repeated because the extensive studies conducted to date have provided no evidence that exposure to ELF EMFs had obvious adverse ecological effects. Although caution must be used in drawing conclusions from the results of most of the studies regarding possible small effects because of faulty study design or analysis, the committee considers it highly unlikely that repetition of the ecological monitoring studies undertaken in this program would produce any new findings about ecological responses to ELF EMFs.

If, in the future, the Navy or another entity endeavors to determine effects, especially small effects, of external factors, such as ELF EMFs, a before-and-after, control-and-impact (BACI) approach is recommended and should be a criterion of the RFP. A BACI approach to the monitoring program would have been appropriate if the antennas had not been functional before many of the studies were initiated. Such an approach would require, however, that the monitoring studies differentiate between the effects due to construction of the external agent and operation of the agent. If future monitoring studies of similar external influences on ecosystem components or processes are planned by the Navy or other entity, hypothesis development should consider an approach used in ecotoxicology: The "null hypothesis" is designated a measurable response (i.e., positive finding) of an ecological attribute to an external influence. The research design is then structured to prove this "null hypothesis" wrong, that is, to prove that there is no measurable response. A finding that results in acceptance of a null hypothesis of no measurable response, as was the case with many of the ELF ecological monitoring studies, does not readily demonstrate that there are no measurable effects, especially small ones. Such a finding could also be due to faulty design or high variation between samples that masked small but real effects of the antenna.

SUGGESTED NEXT STEPS

REANALYSIS OF THE EXPOSURE-ASSESSMENT DATA

The ELF ecological monitoring studies were supplied with ELF-EMF data based on measurements made by IITRI engineering teams. The timing and location of the measurements differed among studies. They included measurements made only once a year at some study plots (as in the wetlands study), at the location of each individual of the response species of interest in others (as in the upland-flora study), and as a spatial gradient of exposure levels (as in the bird-nestling study). In addition to those different forms of available ELF-EMF exposure data, the study teams apparently were made aware of the variability in times and outputs of antenna operations. In some studies, the analyses and interpretations of ELF-EMF effects appear to have made use of the exposure data. However, some studies apparently used the information inappropriately, and others might not have fully recognized the importance of the vagaries of antenna operations and output.

The committee suggests that the investigators from each ecological monitoring study reassess their use of data on antenna operation and ELF-EMF data and, if they were used inappropriately, reanalyze the responses of selected ecological variables. In addition, the results of several studies should be reanalyzed so that firmer conclusions can be drawn. These are the studies labeled in Table 5-1 as "Might Be Acceptable With More Information or Analysis." The committee suggests that an organization that is independent of the U.S. Department of Defense or IITRI manage the reanalysis. The reassessment and reanalysis should be performed in close collaboration with biostatisticians familiar with this type of EMF-exposure assessment and engineers knowledgeable about field ELF-EMF exposure measurements. If reanalysis reveals statistically significant or suggestive responses of ecological variables to ELF EMFs, these responses could be considered for further controlled study (as discussed below). The committee suggests reanalysis and possible controlled studies so that an opportunity is not lost to improve the understanding of ELF-EMF exposure and possible ecological responses.

CONTROLLED LABORATORY STUDIES OF VARIABLES THAT TENDED TO SHOW MEASURABLE EFFECTS

The ELF ecological monitoring studies produced few results that tended to show effects of ELF-EMF exposure on selected ecological variables. One effect, the growth response of upland trees, might be an artifact of selective use of exposure data. In other cases, an effect might be a true measurable response, but the experimental design or the complexity of the surrounding ecosystem might have created an environment that made the findings sufficiently questionable to warrant further, more-controlled studies.

Responses that perhaps could be tested under controlled laboratory conditions are the apparent increase in chlorophyll-a in the aquatic-ecosystem study, the behavioral responses of bees and their overwintering mortality in the pollinating-insects study, and increased moss growth in the litter-decomposition study. The chlorophyll-a increase appeared to be an increase in cell density rather than in chlorophyll per cell; this possibility could be tested and the ecological implications analyzed. A similar study might help in understanding whether alterations in bee behavior and mortality are repeatable and can be shown to be caused by ELF-EMF exposure or are artifacts of the less-controlled, more-complex study sites. The wetland study unexpectedly discovered more moss cover on decomposition bags closer to the antenna treatment sites than in intermediate treatment or background control sites. The increased moss cover caused problems in interpreting data on decomposition, but the variability in growth of moss should be considered for controlled investigation. The reanalysis of exposure assessments (as discussed above) might uncover additional suggestions of small but measurable responses of ecological attributes to ELF-EMF exposure. If it does, these responses could also receive further study under more-controlled conditions; such studies could be designed also to help to elucidate the mechanisms of response if an EMF effect is observed. Such information might guide researchers in deciding which organisms and response variables are most likely to exhibit effects, if any, of the ELF antenna.

ANY REANALYSIS OR LABORATORY STUDIES SHOULD BE REVIEWED INDEPENDENTLY

Reanalysis of exposure assessments might or might not identify some effects of ELF-EMF exposure on ecological variables not previously observed, and laboratory tests might or might not confirm them. Reanalysis might also strengthen the credibility of the findings of some studies. The committee suggests that if reanalyses or laboratory studies are performed, the Navy should arrange for an independent evaluation by a few individuals to assess all of the findings resulting from the reanalysis. The individuals should include biostatisticians familiar with ELF-EMF exposure assessment and biologic expertise to determine what the weight of evidence indicates and the biologic or ecological implications of any substantiated treatment effects. A broader integration of all studies should be pursued through the use of quantitative methods designed for such purposes. Integration of related effects, although not statistically significant, can point to areas where additional study might be warranted. The results of the independent evaluation should be made publicly available. Such an independent final review would serve the Navy and the public in producing more-credible and improved findings of the monitoring program.

SUMMARY

The committee considers it unlikely that further field-based ecological studies will yield more-substantial conclusions than have been obtained from the present studies. Possible weak effects (as in chlorophyll-a, bees, and moss) were identified in the present studies. The committee suggests that some data be reanalyzed to establish whether they can add to confidence in the conclusions of the more acceptable studies or identify some other possible effects. Any studies of such possible effects could be pursued under controlled laboratory conditions and combined with theoretical studies designed to elucidate possible mechanisms. If these strategies are pursued and data reanalysis and laboratory studies are completed, the results of the entire program should be evaluated independently and the results of the evaluation should be made publicly available.

References

AIBS (American Institute of Biological Sciences). 1985. Assessments and Viewpoints on the Biological and Human Health Effects of Extremely Low Frequency (ELF) Electromagnetic Fields. Compilation of Commissioned Papers for the ELF Literature Review Project. Washington, D.C.: American Institute of Biological Sciences.

Anderson, L.E. 1990. Biological effects of extremely low-frequency and 60 Hz fields. Chapter 9 in Biological Effects and Medical Applications of Electromagnetic Energy, O.P. Gandhi, ed. Englewood Cliffs, N.J.: Prentice-Hall.

Antai, S.P., and D.L. Crawford. 1981. Degradation of softwood, hardwood, and grass lignocelluloses by two *Streptomyces* strains. Appl. Environ. Microbiol. 42:378-380.

Beaver, D.L., R.W. Hill, and S.D. Hill. 1994. ELF Communication System Ecological Monitoring Program: Small Vertebrate Studies—Final Report. Technical Report D06212-1. Prepared for Submarine Communications Project Office, Space and Naval Warfare Systems Command, Washington, D.C., by IIT Research Institute, Chicago, Ill.

Brayman, A.A., M.W. Miller, C. Cox, E.L. Carstensen, and M. Schaedle. 1985. Absence of a 45 or 60 Hz electric field-induced respiratory effect in *Physarum polycephalum*. Radiat. Res. 104:242-261.

Bruhn, J.N., S.T. Bagley, and J.B. Pickens. 1994. ELF Communication System Ecological Monitoring Program: Litter Decomposition and Microflora Studies—Final Report. IITRI Tech. Rep. D06212-3. IIT Research Institute, Chicago, Ill.

Burton, T.M., R.J. Stout, S. Winterstein, T. Coon, D. Novinger, R. Stelzer, and M. Rondinelli. 1994. ELF Communications System Ecological Monitoring Program:

REFERENCES

Aquatic Ecosystem Studies—Final Report. IITRI Tech. Rep. D06212-5. IIT Research Institute, Chicago, Ill.

Carpenter, S.R., T.M. Frost, J.F. Kitchell, and T.K. Kratz. 1993. Species dynamics and global environmental change. Pp. 267-279 in Biotic Interactions and Global Change, P.M. Kareiva, J.G. Kingsolver, and R.B. Huey, eds. Sunderland, Mass.: Sinauer Associates.

Caswell, H. 1976. Community structure—Neutral model analysis. Ecol. Monogr. 46(3):327-354.

Crawford, D.L. 1978. Lignocellulose decomposition by selected streptomyces strains. Appl. Environ. Microbiol. 35(6):1041-1045.

Dill, M.W. 1984. ELF ground design and installation. IEEE J. Oceanic Eng. OE-9:128-135.

Goodman, E.M., and B. Greenebaum. 1988. The Effects of Exposing the Slime Mold *Physarum polycephalum* to Electromagnetic Fields, Vol. 1 of 3, Tab C in Compilation of 1987 Annual Reports of the Navy ELF Communications System Ecological Monitoring Program. IITRI Tech. Rep. E06595-2. IIT Research Institute, Chicago, Ill.

Goodman E., and B. Greenebaum. 1990. ELF Communications System Ecological Monitoring Program: Slime Mold Studies—Final Report. IITRI Tech. Rep. E06620-3. IIT Research Institute, Chicago, Ill.

Guntenspergen, G., J. Keough, F. Stearns, D. Wikum. 1988. ELF Communications System Ecological Monitoring Program: Wetlands Studies, Annual Report 1987. Vol. 3 of 3, Tab I in Compilation of 1987 Annual Reports of the Navy ELF Communications System Ecological Monitoring Program. IITRI Tech. Rep. E06595-2. IIT Research Institute, Chicago, Ill.

Guntenspergen, G., J. Keough, F. Stearns, and D. Wikum. 1989. ELF Communications System Ecological Monitoring Program: Wetlands Studies—Final Report. IITRI Tech. Rep. E06620-2. IIT Research Institute, Chicago, Ill.

Hanowski, J., J.G. Blake, G.J. Niemi, and P.T. Collins. 1991. ELF Communications System Ecological Monitoring Program: Wisconsin Bird Studies—Final Report. IITRI Tech. Rep. E06628-2. IIT Research Institute, Chicago, Ill.

Hanowski, J., G.J. Niemi, and J.G. Blake. 1994. ELF Communications System Ecological Monitoring Program: Michigan Bird Studies—Final Report. IITRI Tech. Rep. D06212-2. IIT Research Institute, Chicago, Ill.

Haradem, D.P., J.R. Gauger, and J.E. Zapotosky. 1992. ELF Communications System Ecological Monitoring Program: Electromagnetic Field Measurements and Engineering Support—1992. IITRI Tech. Rep. D06205-1. IIT Research Institute, Chicago, Ill.

Haradem, D.P., J.R. Gauger, and J.E. Zapotosky. 1994. ELF Communications System Ecological Monitoring Program: Electromagnetic Field Measurements and Engineering Support—Final Report. IITRI Tech. Rep. D06209-1. IIT Research Institute, Chicago, Ill.

Hurlbert, S.H. 1984. Pseudoreplication and the design of ecological field experiments. Ecol. Monog. 54(2):187-211.
IITRI (Illinois Institute of Technology Research Institute). 1976. Seafarer Monitoring Plan. Prepared by Equitable Environmental Health, Rockville, Md., for IIT Research Institute, Chicago, Ill.
Jones, E.A., D.D. Reed, P.J. Cattelino, and G.D. Mroz. 1991. Seasonal shoot growth of planted red pine predicted from air-temperature degree days and soil-water potential. For. Ecol. Manage. 46(3-4):201-214.
Knutson, D.M., A.S. Hutchins, and K. Cromack. 1980. The association of calcium oxalate utilizing Streptomyces with conifer ectomycorrhizae. Antonie van Leeuwenhoek 46(6):611-619.
Krizaj, D., and V. Valencic. 1989. The effect of ELF magnetic-fields and temperature on differential plant-growth. J. Bioelectricity 8(2):159-165.
Martin, A.H. 1989. Electromagnetic fields and the developing embryo [abstract]. P. A-15 in the Annual Review of Research and Biological Effects of 50- and 60-Hz Electric and Magnetic Fields, sponsored by the U.S. Department of Energy, Nov. 13-16, 1989, Portland, Oreg.
Marx, D.H. 1982. Mycorrhizae in interactions with other microorganisms. Pp. 225-228 in Methods and Principles of Mycorrhizal Research. N.C. Schenck, ed. St. Paul, Minn.: American Phytopathological Society.
McClaugherty, C.A., J. Pastor, J.D. Aber, and J.M. Melillo. 1985. Forest litter decomposition in relation to soil nitrogen dynamics and litter quality. Ecology 66(1):266-275.
Meentemeyer, V. 1978. Macroclimate and lignin control of litter decomposition rates. Ecology 59(3):465-472.
Melillo, J.M., J.D. Aber, and J.M. Muratore. 1982. Nitrogen and lignin control of hardwood leaf litter decomposition dynamics. Ecology 63(3):621-626.
Melillo, J.M., R.J. Naiman, J.D. Aber, and A.E. Linkens. 1984. Factors controlling mass-loss and nitrogen dynamics of plant litter decaying in northern streams. Bull. Mar. Sci. 35:(3) 341-356.
Michaelson, S.M. and J.C. Lin. 1987. Pp. 157-164 in Biological Effects and Health Implications of Radiofrequency Radiation. New York: Plenum Press.
Mroz, G.D., P.J. Cattelino, M.R. Gale, E.A. Jones, M.F. Jürgensen, H.O. Liechty, and D.D. Reed. 1994. ELF Communications System Ecological Monitoring Program: Upland Flora Studies—Final Report. Tech. Rep. D06212-4. IIT Research Institute, Chicago, Ill.
Nelson, S.O. 1991. Dielectric-properties of agricultural products—Measurements and applications. IEEE Trans. Electr. Insul. 26(5):845-869.
Niemi, G.J., and J.M. Hanowski. 1984. A Proposal for a Study to Monitor the Effects of the ELF Antenna System on Bird Species and Communities: 1985-1987. Submitted to IIT Research Institute. Chicago, Ill.
NIOSH (National Institute for Occupational Safety and Health), NIEHS (National

Institute of Environmental Health Sciences, and DOE (U.S. Department of Energy). 1996. Questions and Answers: EMF in the Workplace. Washington, D.C.: Superintendent of Documents, U.S. Government Printing Office.

NRC (National Research Council). 1977. Biologic Effects of Electric and Magnetic Fields Associated with Proposed Project Seafarer. Assembly of Life Sciences. Washington, D.C.: National Academy of Sciences.

NRC (National Research Council). 1997. Possible Health Effects of Exposure to Residential Electric and Magnetic Fields. Washington, D.C.: National Academy Press.

NRPB (National Radiological Protection Board). 1992. Electromagnetic Fields and the Risk of Cancer, Vol. 3. Chilton, Didcot, U.K.: National Radiological Protection Board.

Olson, J.S. 1963. Energy storage and the balance of producers and decomposers in ecological systems. Ecology 44(2):322-331.

ORAU (Oak Ridge Associated Universities) Panel. 1992. Health Effects of Low-Frequency Electric and Magnetic Fields. ORAU 92/F8. Committee on Interagency Radiation Research and Policy Coordination, Washington, D.C.

ORAU (Oak Ridge Associated Universities) Panel. 1993. EMF and cancer. Science 260:13-14.

OTA (Office of Technology Assessment). 1989. Biological Effects of Power Frequency Electric and Magnetic Fields—Background Paper. U.S. Congress, Office of Technology Assessment. OTA-BP-E-53. Washington, D.C.: U.S. Government Printing Office.

Pastor, J., and W.M. Post. 1986. Influence of climate, soil moisture, and succession on forest carbon and nitrogen cycles. Biogeochemistry 2:03-27.

Polk, C., and E. Postow. 1986. CRC Handbook of Biological Effects of Electromagnetic Fields. Boca Raton, Fla.: CRC Press.

Prosser, C.L. 1973. Comparative Animal Physiology, 3rd Ed. Philadelphia, Pa.: Saunders.

Reed, D.D., E.A. Jones, M.J. Holmes, and L.G. Fuller. 1992. Modeling diameter growth in local-populations: A case-study involving four North-American deciduous species. For. Ecol. Manage. 54(1-4):95-114.

Reed, D.D., E.A. Jones, G.D. Mroz, H.O. Liechty, P.J. Cattelino, and M.F. Jürgensen. 1993. Effects of 76 Hz electromagnetic-fields on forest ecosystems in northern Michigan—Tree growth. Int. J. Biometerol. 37(4):229-234.

Richter, D.L., T.R. Zuellig, S.T. Bagley, and J.N. Bruhn. 1989. Effects of red pine (pinus-resinosa Ait.) mycorrhizoplane-associated actinomycetes on in vitro growth of ectomycorrhizal fungi. Plant and Soil 115(1):109-116.

Smith, M.L., J.N. Bruhn, and J.B. Anderson. 1992. The fungus Armillaria bulbosa is among the largest and oldest living organisms. Nature 356:428-431.

Stearns, F., J. Keough, N.P. Lasca, and C.-Y. Yuen. 1982. Ecology and Geology of the Superior Upland Region: A Theme Study for the National Park Service. University of Wisconsin-Milwaukee.

Stearns, F., G. Guntenspergen, and J. Keough. 1984. ELF Communications System Ecological Monitoring Program: Wetland Studies, Annual Report, 1983. Vol. 2 of 2, Tab I, in Complication of 1983 Annual Reports of the Navy ELF Communications System Ecological Monitoring Program. IITRI Tech. Rep. E06549-8. IIT Research Institute, Chicago, Ill.

Strickler, K., and J.M. Scriber. 1994. ELF Communications System Ecological Monitoring Program: Pollinating Insect Studies—Final Report. IITRI Tech. Rep. D06212-6. IIT Research Institute, Chicago, Ill.

Tenforde, T. S. 1996. Interaction of ELF magnetic fields with living systems. Pp. 185-230 in Handbook of Biological Effects of Electromagnetic Fields, 2nd Ed. C. Polk and E. Postow, eds. Boca Raton, Fla.: CRC Press.

Wiewiorka, Z. 1990. The effects of electromagnetic and electrostatic fields on the development and yield of greenhouse tomato plants. Acta Agrobot. 43(1-2):25-36.

Wiewiorka, Z., and J. Sarosiek. 1987. The effects of nononizing radiation on the aquatic liverwort ricciocarpus-natans 1. corda. Symp. Biol. Hung. 15:849-856.

Wilson, B.W., R.G. Stevens, and L.E. Anderson. 1989. Neuroendocrine mediated effects of electromagnetic-field exposure: Possible role of the pineal gland. Life Sci. 45(15):1319-1332.

Zapotosky, J.E., and J.R. Gauger. 1993. ELF Communications System Ecological Monitoring Program—Summary Report for 1982-1992. IITRI Tech. Rep. D06205-3. IIT Research Institute, Chicago, Ill.

Zapotosky, J.E., J.R. Gauger, and D.P. Haradem. 1996. ELF Communications System Ecological Monitoring Program—Final Summary Report. IITRI Tech. Rep. D06214-6. IIT Research Institute, Chicago, Ill.

Appendix A

Biographic Information on Committee Members

DUNCAN PATTEN (CHAIR) is professor emeritus of botany and past director of the Center for Environmental Studies at Arizona State University. He is also research professor with the Mountain Research Center at Montana State University. Dr. Patten has an AB from Amherst College, an MS from the University of Massachusetts at Amherst, and a PhD from Duke University. His research interests include arid and mountain ecosystems, especially ecological processes of western riparian and wetland ecosystems. He has been senior scientist of the Bureau of Reclamation's Glen Canyon environmental studies, overseeing the research program that evaluates the effects of operations of Glen Canyon Dam. Dr. Patten also served as business manager for the Ecological Society of America for 16 years and is president of the Society of Wetland Scientists. At the National Research Council, he has been a member of the Commission on Geosciences, Environment, and Resources, the Board on Environmental Studies and Toxicology, and numerous committees.

OM P. GANDHI is professor and chairman of the Department of Electrical Engineering at the University of Utah, Salt Lake City. He is author or coauthor of book chapters and journal articles on electromagnetic dosimetry, microwave tubes, and solid-state devices; editor of *Biological Effects and Medical Applications of Electromagnetic Energy* (Prentice-Hall, 1990); and coeditor of *Electromagnetic Biointeraction* (Plenum Press, 1989). Dr. Gandhi was elected a fellow of the Institute of Electrical and Electronics Engineers (IEEE) in 1979 and received the Distinguished Research Award from the University of Utah for 1979-1980. He has been president of the Bioelectromagnetics Society (1992-1993), cochairman of the IEEE SCC 28.IV Subcommittee on RF Safety Standards (1988-present), and chairman of the IEEE Committee on Man and Radiation (1980-1982). In 1995, he received the d'Arsonval Medal of the Bioelectromagnetics Society for pioneering contributions to the field.

THOMAS GETTY is associate professor at the Kellogg Biological Station (KBS) and the Department of Zoology, Michigan State University. He is a principal investigator in the National Science Foundation Research Training Group: Linking Levels of Ecological Organization at KBS. He teaches ecology, behavior, and mathematical modeling and does research on how animals adapt to variation and uncertainty in their environments. He has a BS in engineering and a PhD in biology from the University of Michigan.

WILLIAM E. GORDON is a consulting engineer in wireless communication and remote sensing. During the first half of his academic career (1948-1966), at Cornell University, he conceived, supervised the design and construction of, and directed the early operation of the Arecibo Observatory, which has a 300-m spherical antenna. At Rice University (1966-1986), he served as a professor of space science and electrical engineering, as dean of sciences and engineering, as provost and as vice president; he is now a distinguished professor emeritus. He is a member of the National Academy of Sciences (and was foreign secretary in 1986-1990), a member of the National Academy of Engineering, a foreign associate of the Engineering Academy of Japan, and a fellow of the American Academy of Arts and Sciences, the American Association for the Advancement of Science, the American Geophysical Union, and the Institute of Electrical and Electronic Engineers. He was vice president of the International Council of Scientific Unions (1988-1993) and is an honorary president of the International Union of Radio Science. He received the Balth van der Pol Gold Medal in 1966, the Arctowski Gold Medal in 1984, a USSR Academy of Sciences Medal in 1985 for distinguished contributions in international geophysical programs, and the Centennial Medal of the University of Sofia in 1988.

J. WOODLAND HASTINGS is Paul C. Mangelsdorf Professor of Natural Sciences in the Department of Molecular and Cellular Biology, Harvard University. He was previously on the faculty at the University of Illinois, Urbana and Northwestern University, Evanston, and a visiting professor at Rockefeller University. His research is concerned with cellular and biochemical mechanisms of biologic oscillations, particularly circadian rhythms. He is also an authority on the biochemistry and physiology of bioluminescence in organisms ranging from bacteria to vertebrates. He served as chairman of the 1976 National Research Council Committee on Biologic Effects of Electric and Magnetic Fields Associated with Proposed Project Seafarer. A graduate of Swarthmore College, Dr. Hastings obtained a PhD in biology from Princeton University.

PETER KAREIVA is professor in the Department of Zoology, University of Washington in Seattle. He obtained a BS from Duke University, an MS from the University of California, Irvine, and a PhD in ecology and evolution from Cornell University. He has been a member of the Ecological Society of America and the Entomological Society of America. His research interests include population biology of herbivorous insects, mathematical models of insect dispersal, and the influence of vegetation texture on herbivore dynamics.

JAMES C. LIN is professor of electrical engineering, bioengineering, physiology and biophysics at the University of Illinois, Chicago, where he also served as head of the Department of Bioengineering in 1980-1992. He chairs the National Council on Radiation Protection and Measurements, the Scientific Committee on Biological Effects and Exposure Criteria for Radio Frequency Fields, and the US National Committee for Radio Science, Commission on Electromagnetics in Biology and Medicine. Dr. Lin holds a research chair from the National Science Council. He was president of the Bioelectromagnetics Society, chairman of the Institute of Electrical and Electronics Engineers (IEEE) Committee on Man and Radiation, and president of the Chinese American Academic and Professional Association. He is a Fellow of the American Association for the Advancement of Science, the American Institute for Medical and Biological Engineering, and IEEE. He has published over 100 papers in refereed journals and contributed to 15 book chapters. He has written two books and edited two books. He received a BS (1966), an MS (1968), and a PhD (1971) in electrical engineering from the University of Washington, Seattle.

ROBERT G. OLSEN is professor of electrical engineering and computer science at Washington State University. He has held positions with Westinghouse Georesearch Laboratory, GTE Laboratories, and ASEA Research Laboratory in Sweden. He has been a member of the Technical Committee of the Washington State EMF Task Force and chair of the Institute of Electrical and Electronics Engineers (IEEE) Power Engineering Society AC Fields Working Group. He is now chair of the IEEE Power Engineering Society Corona Effects Working Group and an associate editor of the IEEE *Transactions on Electromagnetic Compatibility*. He is a fellow of the IEEE. He received a BS in Electrical Engineering from Rutgers University in 1968 and an MS and a PhD from the University of Colorado, Boulder in 1970 and 1974, respectively.

JOHN PASTOR is professor of biology and senior research associate at the Department of Biology and the Natural Resources Research Institute, University of Minnesota, Duluth. In addition, he is adjunct professor in the Department of Ecology and Behavioral Biology, University of Minnesota-Minneapolis and the Department of Fisheries and Wildlife, University of Minnesota-St. Paul. He received a BS in geology from the University of Pennsylvania, and an MS in soil science and a PhD in forestry and soil science from the University of Wisconsin-Madison.

BEVERLY J. RATHCKE is associate professor in the Department of Biology at the University of Michigan. She received a BA from Gustavus Adolphus College, an MSc from the University of London, and a PhD from the University of Illinois at Urbana-Champaign. She was a Fulbright fellow at the University of London, and a NATO post-doctoral fellow at the University College of North Wales and has held research positions at Brown University and Cornell University. She was an editor for the Ecological Society of America. Her research interests are community ecology and plant-animal interactions.

ANTONIO SASTRE is Principal Scientist at the Health Assessment and Research Center at the Midwest Research Institute in Kansas City, Missouri. He has been on the full-time faculty at the Johns Hopkins University School of Medicine in physiology (1977-1988) and neuroscience (1980-1986) at the Assistant and Associate Professor level. He is also Adjunct Associate Professor of Pharmacology at the Cornell University Medical College (1979-1996). Dr. Sastre received his B.A. in 1970, an M.S. in 1973 and a Ph.D. in applied mathematics in 1974 from Cornell University. His areas of research include systemic and cellular pharmacology, cardiovascular physiology and neurobiology; membrane and receptor biophysics; digital processing and modeling of bioelectric signals; bioelectromagnetics research on in vivo responses of human subjects and biophysical modeling of cell responses.

LAWRENCE A. SHEPP is a mathematician who has been at Bell Laboratories for 34 years and is a member of the National Academy of Sciences and the Institute of Medicine. He is now teaching at Columbia University in the Departments of Statistics and Operations Research. He has extensive engineering experience in computed axial tomography scanning and magnetic-resonance imaging.

Appendix B

Calculations of Induced Electric Fields

INTRODUCTION

THIS APPENDIX PROVIDES calculations on the relative strengths of induced electric fields in various biota exposed to 76-Hz electric and magnetic fields.

THEORETICAL MODEL

A number of investigations have used tissue-equivalent spheroidal models as an index of induced field. The spheroidal model is attractive because simple expressions can be obtained for all body sizes. The wavelength at 76 Hz is very large, compared with the longest dimension of the body, so the quasi-static field theory can be appropriately applied to calculate the induced electric field in the body (Michaelson and Lin 1987). For uniform external electric and magnetic fields, the magnitude of the induced electric field inside a homogeneous dielectric tissue sphere resulting from the applied electric field is

$$E_e = 3/\epsilon E_0 \tag{B-1}$$

and the peak magnitude of the induced electric field resulting from the applied magnetic field is

$$E_m = \pi a f B_0 \tag{B-2}$$

where ϵ is the dielectric permittivity, a is the radius, f is the frequency, E_0 is the imposed or applied electric-field strength, and B_0 is the magnetic flux density. The uniform external electric field gives rise to a constant induced electric field inside the dielectric sphere that has the same direction but is reduced by $3/\epsilon$ from the applied electric field for the biologic object and is independent of body size. The magnetically induced electric field produces an internal electric field that varies directly with the radius of the spherical body and is proportional to the source frequency.

For some species, a prolate spheroidal model approximates more closely their elongated bodies. The magnitude of the electric field induced inside a homogeneous dielectric spheroid with semimajor axis a and semiminor axis b by a uniform applied electric field oriented along the semimajor axis is

$$E_{ee} = E_0/C_1 \tag{B-3}$$

and for an electric field oriented along the semi-minor axis of the body is

$$E_{he} = -E_0/C_2. \tag{B-4}$$

Similarly, the peak magnitude of the electric field induced by a uniform magnetic field oriented along the semimajor axis is

$$E_{em} = \pi b f B_0 \tag{B-5}$$

and for a magnetic field oriented along the semiminor axis of the body is

$$E_{hm} = \pi a f B_0. \tag{B-6}$$

Because C_1 and C_2 are constants, the induced fields are uniform. However, they are dependent on the orientation of the applied electric and magnetic fields with respect to the major axis of the body. In particular, because $a > b$, the higher induced field is associated with an applied magnetic field oriented along the minor axis of the body.

For both spheroidal models, the electrically induced current is in the direction of the applied field and is uniform. The magnetically induced current is a circulating current with an amplitude of zero at the center of the body and increases with distance from the center. In all cases, the electrically induced field is uniform, but the magnetically induced field increases with increasing size of the subject, such as the average radius or longest dimension of the body.

NUMERICAL CALCULATIONS

To provide an index of induced electric fields in biota and a guide to the extrapolation of data from the ecological monitoring program to other experimental subjects, the committee has made numerical calculations of induced electric field as a function of size (1 mg to 500 g), using spheres to approximate the shapes of insects, birds, and small vertebrates (see Equations B-1 and B-2 above and data given in Polk and Postow 1986). In addition, an elongated prolate spheroid (see Equations B-3 through B-6 above and data given in Nelson 1991) is used to model upland hardwood-tree stands whose average diameter is 15-25 cm and average height is 10-20 m. The exposure parameters considered are applied electric fields of 10-5,000 mV/m and applied magnetic fields of 1.0-50 mG. Results are shown in Figures B-1 through B-3.

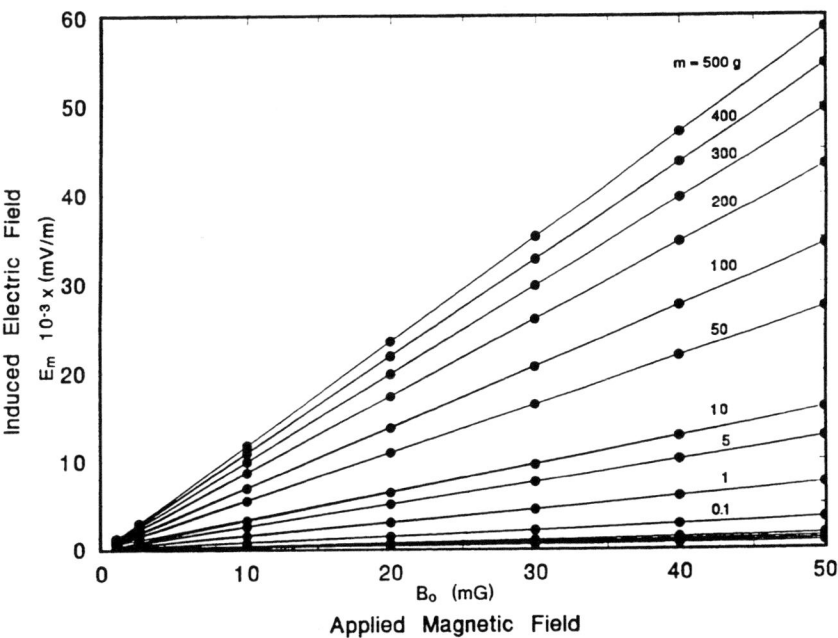

FIGURE B-1 Magnetically induced electric field in insects and birds and other vertebrates.

INSECTS AND BIRDS AND OTHER SMALL VERTEBRATES

It is noted that electrically induced fields are the same for all body sizes and are proportional to the strength of the applied electric fields. For the parameters considered, the values are lower than 16×10^{-5} mV/m and are less than one-millionth of the applied electric fields.

As shown in Figure B-1, the magnetically induced electric field varies with both body size and magnetic field. For a 100-mg insect, the maximal induced electric field can vary from 6.9×10^{-5} mV/m at 1.0 mG to 3.4×10^{-3} mV/m at 50 mG. Likewise, for a 100-g bird or other vertebrate, the highest induced electric field varies from 6.9×10^{-4} mV/m at 1.0 mG to 3.4×10^{-2} mV/m at 50 mG.

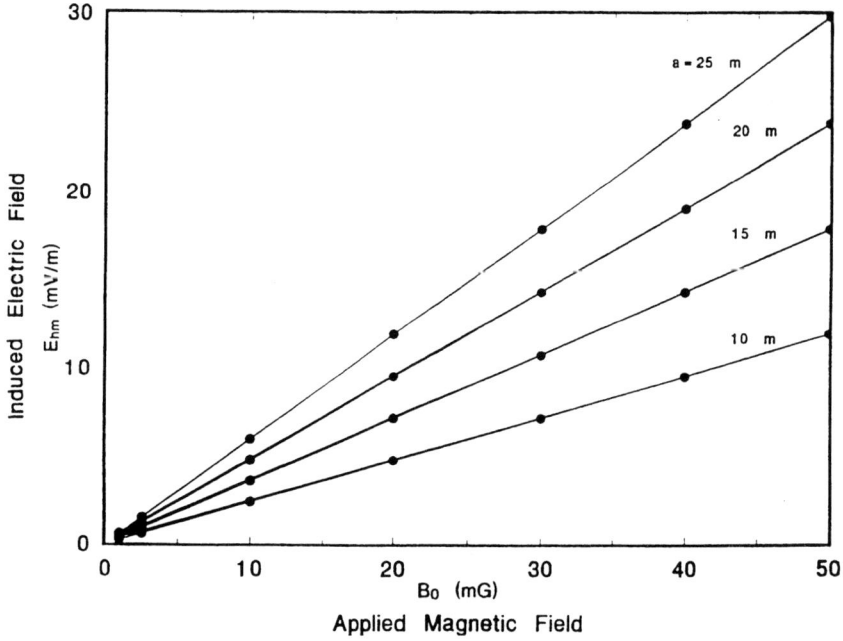

FIGURE B-2 Vertical magnetic-field-induced electric field in hardwood-tree stand.

HARDWOOD-TREE STANDS

The induced electric field resulting from an applied electric field oriented vertically along the height (major axis) of a hardwood stand is independent of stand size. Because the vertical electric field is tangential to the major dimension of the tree stand, the induced and applied electric fields are the same (10-5,000 mV/m).

Although the induced electric field resulting from an applied electric field oriented horizontally along the width (minor axis) of the tree stand is also independent of stand size, the values are drastically reduced and vary from 1.6×10^{-2} to 8.19 mV/m for applied fields of 10-5,000 mV/m.

The results of a vertically oriented magnetic field are shown in Figure B-2. Induced electric fields depend both on the width of the tree stand and on the magnitude of the applied magnetic field. For a 10-cm width, the induced electric fields resulting from applied magnetic fields of 10 and 50 mG are 1.2

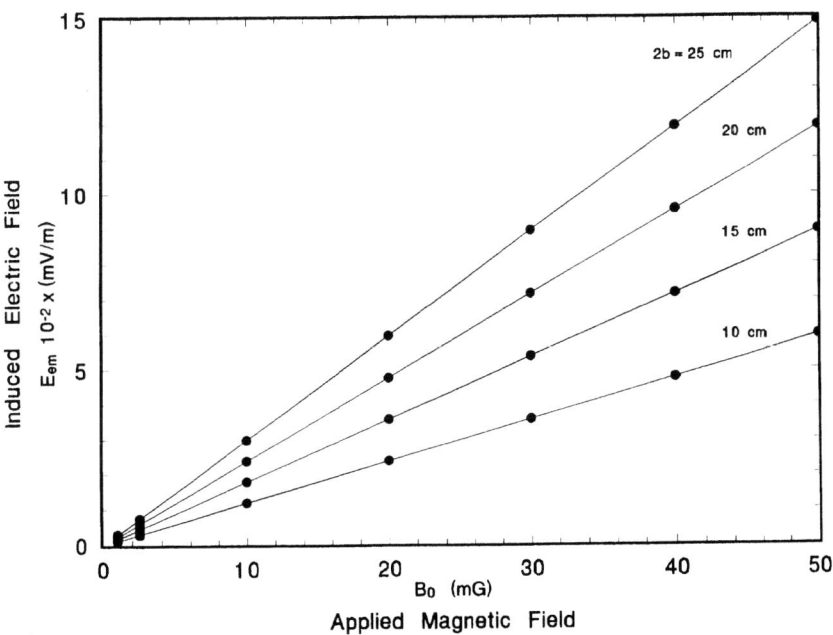

FIGURE B-3 Horizontal magnetic-field-induced electric field in hardwood-tree stand.

× 10^{-2} and 6.0×10^{-2} mV/m, respectively. For these magnetic-field magnitudes, the induced electric fields in a 25-cm tree stand are 3.0×10^{-2} and 0.15 mV/m, respectively.

If the applied magnetic field is oriented horizontally along the minor axis of the tree stand, the induced electric field will be proportional to the height of the tree stand and the strength of the magnetic field. The electric fields induced by magnetic fields of 1.0 and 50 mG in a tree 10 m tall are 0.24 and 11.9 mV/m, respectively, and in a 25-m tree are 0.6 and 29.8 mV/m, respectively (see Figure B-3).

SUMMARY

In summary, because of small size, the calculated 76-Hz electric fields induced in insects, birds, and small vertebrates by electric fields of up to 5,000 mV/m and magnetic fields of up to 50 mG are fairly low. In contrast, electric fields induced by the same electric and magnetic fields in large hardwood-tree stands could be substantial. Calculations based on these simple models suggest that the electric field induced in a 25-m tree by a vertically oriented electric field could be as high as 5,000 mV/m and that induced by a horizontally oriented magnetic field could be as high as 29.8 mV/m. It is emphasized that because of shielding and other phenomena, the applied or impinging electric field would decrease in strength with distance from the antenna wire and as a function of the landscape. However, magnetic-field strength would remain unattenuated by its environment and would decrease in strength only with distance from the antenna wire because magnetic permeability remains unchanged. Therefore, at greater distances from the antenna, the electric field induced in tree stands by a horizontal magnetic field could become a dominant factor.